Preface

During the past 15 to 20 years, a need for deep water structures that would exploit energy resources such as oil and natural gas has arisen. When deep water combined with hostile weather conditions are considered, conventional fixed offshore structures require excessive physical dimension to obtain the needed stiffness and strength, and therefore are very costly. Thus, special deep water platforms called *compliant offshore structures* had to be developed. This kind of structure is flexibly linked to the seafloor and is free to move with the waves. Since the compliant structure cannot resist lateral forces due to waves, current and/or wind, the restoring moment is provided by a large buoyancy force, a set of guylines or a combination of both. The structure's fundamental frequency is designed to be well below the wave lower frequency-bound in order to avoid harmonic resonances. Forces on these structures are due to ocean and atmosphere include ocean waves and current, wind, buoyancy, and friction at the base. This results in relatively large displacements, and thus geometric nonlinearity is an important consideration in the analysis of such structures. Various models have been developed for these forces, with a spectrum of sophistication. Similarly, the complex equations of motion governing structural response find numerous methods for their solution. Four types of structures fall into the category of compliant structures: *guyed towers, tension leg platforms, risers* and *articulated towers*. These structures have found primary offshore application in the oil industry, but also for cases where a stable ocean platform is needed for communication, mooring and even for 'floating' airports. The purpose of this monograph is to provide an analytical mechanics framework for the formulation of the nonlinear equations of motion of compliant structures in the ocean. Such an analytical/computational treatment is viewed as a first step in the overall process of designing and erecting these enormous and wonderful structures. The usefulness of such formulations, albeit with all their simplifications and assumptions, is that they prepare us for possible and perhaps likely dynamic behavior characteristics of the full structure. These models can warn the designer of dangerous regimes of behavior and inappropriate dimensions. Such modeling exercises are a road map for the designers as they prepare preliminary designs, million degree of freedom computational models, and scale models for testing.

The beginning of each chapter summarizes the literature, focusing on articulated towers, tension leg platforms and risers, respectively. Chapter 1 provides the theoretical basis and derivations upon which all the subsequent chapters are based. Chapter 2, by far the longest and most detailed one, discusses the behavior of an articulated tower. Chapter 3, continues with a tension leg platform. Chapter 4 concludes with a riser with internal flow. All these structures have similarities and, of course, differences. A unifying perspective is presented and an extensive list of references is included. This work is not intended as a recap of a completed work, but rather a convenient point at which to stop, gather our understanding, take a deep breath, and continue.

The numerical results presented in this monograph were created using the following commercially available codes: MAPLE, MATLAB, ACSL, in conjunction with Fortran. The codes were developed by the first author, and are available in printed form upon written request by letter to the following address: Dr. Patrick Bar-Avi, P.O. Box 2250, Dept. 4P, Haifa, Israel, 31021. There is no warrantee, implied or otherwise, in the use of these codes. The final judgement is that of the user of these codes.

Acknowledgments

We would very much like to acknowledge the numerous people and organizations that created this opportunity for us to work on such interesting problems. The first author thanks his sponsoring organization, RAFAEL - Armament Development Authority of Israel, which has permitted him to take several years away from his responsibilities at work, and to partially support him, to pursue the Ph.D. at Rutgers University. Both authors wish to thank Dr. Thomas Swean, and the Office of Naval Research, for sponsoring portions of this work under Grant No. N00014-93-1-0763. We are deeply grateful. Both authors also are pleased to thank the Department of Mechanical and Aerospace Engineering and the College of Engineering, at Rutgers for the comfortable environment within which this work was accomplished.

We want to thank Professor Fai Ma for his encouragement and Martin Scrivener for his publishing support.

We are also grateful to the following colleagues and friends who agreed to read though all or portions of this manuscript, looking furiously for dangling subscripts, missing derivatives or exponents, as well as mixed grammatical usage: Ron Adrezin, Diana Melick and Mindy Schretter.

Finally the authors would like to thank the following software companies, 'MGA', 'Math Works', 'Maple' and 'TCI', for providing us with the tools that were crucial for this work.

1
Theoretical Background: The General Model

This chapter provides the theoretical background and basis for the specific applications of the subsequent chapters. The first section will provide a brief introduction to the *linear wave theory* that forms the basis for the derivation of the wave forces on the structure. Since this monograph considers only slender members, diffraction theory is not discussed. Once the basic wave hydrodynamics is formulated, it is possible to proceed with the approach taken here to estimate wave forces on slender members, that is, the *Morison equation*, arguably the single most important equation for such applications and now over fifty years old. Finally in this chapter, a general derivation of the structural equations of motion is detailed. These governing equations are the starting point for the applications that follow. To make this derivation of broadest use, more details are provided than is customary in technical papers, thus, making this book a working partner of the reader/designer as the contents are applied further.

Linear Wave Theory

There are numerous books on fluid dynamics and ocean engineering that derive the equations for small amplitude gravity wave theory; see for example [1] . For completeness, such a derivation is provided here. Only equations that are required later will be derived and discussed. Such a wave theory is really only a first approximation to the actual behavior of ocean waves. But this simplest model provides much understanding of the fluid-structure interaction, while excluding phenomena that result from large amplitude, nonlinear waves. Only two dimensional motion is considered;

see Fig. 1.1. In the discussion that follows, standard notation is used. When applied to the specific problems of the next three chapters, local notation is adopted.

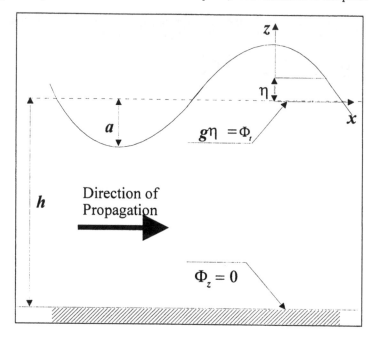

1.1 Schematic of Gravity Waves.

From elementary fluid mechanics, the existence of a velocity potential is assumed, which implies constant density, frictionless, irrotational flow. Assuming small amplitudes permits the elimination of nonlinear terms and results in the general equation of motion of water waves,

$$-\frac{\partial \Phi}{\partial t} + \frac{p}{\rho} + gz = 0,\tag{1.1}$$

where Φ is the velocity potential, p is the fluid pressure, ρ is the fluid density, z is the wave height measured positive upwards from the mean water level and g is the gravitational constant. For a body of water with mean water level h, the boundary value problem to be solved is given by

$$\frac{\partial^2 \Phi}{\partial x^2} + \frac{\partial^2 \Phi}{\partial z^2} = 0,\tag{1.2}$$

where the boundary conditions are given by

$$\left[\frac{\partial \Phi}{\partial z}\right]_{z=-h} = 0, \qquad g\eta = \left[\frac{\partial \Phi}{\partial t}\right]_{z=0},\tag{1.3}$$

where η is the wave elevation (surface profile) given by

$$\eta = a \sin(kx - \omega t),\tag{1.4}$$

which is periodic in x and t, and a is the amplitude (see Fig. 1.1). The wave angular frequency, ω, and the wave number, k, are related as follows

$$\omega^2 = gk \tan kh. \tag{1.5}$$

Partial differential equation (1.2) is solved by the method of separation of variables, resulting in the equation for the velocity potential

$$\Phi = \frac{ag \cosh k(h+z)}{\omega \cosh kh} \cos(kx - \omega t), \tag{1.6}$$

Equation (1.6) is used to derive the horizontal and vertical components of the local fluid particle velocity:

$$u = -\frac{\partial \Phi}{\partial x} = \frac{agk \cosh k(h+z)}{\omega \cosh kh} \sin(kx - \omega t) \tag{1.7}$$

$$w = -\frac{\partial \Phi}{\partial z} = -\frac{agk \sinh k(h+z)}{\omega \cosh kh} \cos(kx - \omega t). \tag{1.8}$$

These equations express the velocity components within the wave at any depth z. For small amplitude waves, the vertical velocity component can also be evaluated as $\partial \eta / \partial t \approx w$,

$$w = \frac{d\eta}{dt} = \frac{\partial \eta}{\partial t} + \left[\frac{\partial \eta}{\partial x} \frac{\partial x}{\partial t} \right], \tag{1.9}$$

where the term in [ ] is neglected. This is the *convective* term.

The local accelerations are straighforwardly obtained by differentiating u and w with respect to time. As before, the convective terms are ignored,

$$\frac{\partial u}{\partial t} = -\frac{agk \cosh k(h+z)}{\cosh kh} \cos(kx - \omega t) \tag{1.10}$$

$$\frac{\partial w}{\partial t} = -\frac{agk \sinh k(h+z)}{\cosh kh} \sin(kx - \omega t). \tag{1.11}$$

For "deep water", the above equations simplify as follows

$$\left. \begin{array}{l} \omega^2 = kg \\ \frac{\cosh kz}{\sinh kh} = e^{kh} \\ \frac{\sinh kz}{\sinh kh} = e^{kh} \end{array} \right\}$$

Thus, the velocities are

$$u = \frac{agk}{\omega} e^{kh} \sin(kx - \omega t) \tag{1.12}$$

$$w = -\frac{agk}{\omega} e^{kh} \cos(kx - \omega t), \tag{1.13}$$

and the accelerations

$$\dot{u} = -agke^{kh} \cos(kx - \omega t) \tag{1.14}$$

$$\dot{w} = -agke^{kh} \sin(kx - \omega t). \tag{1.15}$$

Applied External Forces

Deterministic Model

    Analytical representations of the fluid-structure interaction are complicated by many phenomena, such as turbulent flow about the structure, and the nonlinear relation between water particle kinematics and water surface displacement. The shedding of von Karman vortices leads to structural vibration in a direction transverse to fluid flow, and multiple resonances are possible due to the broad band loadings. In this analysis a circular cylinder is considered since it is the most widely used structural component.

    The wave force on an object immersed in the ocean is equal to the sum of the drag and inertia components of the fluid-structure interaction forces. The purpose is to relate these forces to the wave kinematics. The force is physically the net resultant of the pressure and shear distribution on the cylinder surface. The complex flow field about the cylinder, including wake effects that are significant in vortex shedding, leads the community of researchers to *semi-empirical* relations for the drag and inertia force components. For a *rigid* structure, the drag and inertia forces for an element of length ds is

$$dF_D \;\; = \;\; C_D \rho D \frac{u \mid u \mid}{2} ds \qquad\qquad (1.16)$$

$$dF_M \;\; = \;\; C_M \rho \frac{\pi D^2}{4} \frac{du}{dt} ds, \qquad\qquad (1.17)$$

where ρ is the mass density of water, u is the instantaneous water particle velocity, du/dt is the instantaneous water particle acceleration, both in a direction normal to the cylinder. C_D is the "drag" coefficient, and C_M is the "inertia" or "mass" coefficient. These coefficients vary as a function of *Reynolds number,* but are essentially constant over a large range. Where structural vibration is included, u is replaced by the relative velocity between structure and fluid, and du/dt is replaced by the relative acceleration between structure and fluid.

    The sum

$$dF = C_D \rho D \frac{u \mid u \mid}{2} ds + C_M \rho \frac{\pi D^2}{4} \frac{du}{dt} ds \qquad\qquad (1.18)$$

is known as the *Morison equation,* after the name of the first author of the celebrated paper that first provided a semi-empirical approach to estimating such forces [2] .

Random Model

    As anyone who has observed the ocean knows quite well, there is very little that is exactly defined about the ocean waves. They behave erratically, where the sequence of wave heights are less predictable than that of the outcomes of a series of coin tosses. The analogy turns out to be a reasonable one since in a statistical sense some reproducible behavior is observed. Predictions on the statistical behavior of the wave are made, resulting in probabilistic models for the wave behavior which is then used in the development of random force models. The random model for wave force is formally the same as equation 1.18, except that the fluid kinematics, u, du/dt, are now assumed to be random processes governed by a power spectrum. There are numerous such spectra depending on the particular site for which the wave heights are needed. A commonly used first spectrum is the *Pierson-Moskowitz* ocean wave height spectrum, which will be used in a subsequent chapter. The power spectrum is a measure of the energy in a dynamic process, and is in the form of an energy distribution as a function of frequency. The Morison equation is the key equation for both deterministic and random models of the wave force.

Synthesis of Time Histories from Power Spectra: For simulation of the response in the time domain, the wave height power spectrum is transformed into a time history. This is accomplished using a method by Borgman [3], also described in Wilson's book [4].

The wave elevation $\eta(x,t)$ can be expressed as

$$\eta(x,t) = \int_0^\infty \cos(kx - \omega t + \varepsilon)\sqrt{A^2(\omega)d\omega}, \qquad (1.19)$$

where $A^2(\omega)$ is the amplitude spectrum (height) and ε is a random phase angle having a uniform distribution over an interval 0 to 2π. To evaluate the integral, the spectrum is discretized into equal areas (not equal frequencies). This procedure avoids the presence of periodicities in the resulting time history.

Consider the following partition ;

$$\omega_0 < \omega_1 < \omega_2 < .... < \omega_N = F, \qquad (1.20)$$

where ω_0 is a small positive value and F is the frequency above which the spectral amplitude is practically zero. Let

$$\Delta\omega_n = \omega_n - \omega_{n-1} \qquad (1.21)$$

$$\bar{\omega}_n = \frac{\omega_n + \omega_{n-1}}{2}, \qquad n = 1, 2, ..., N, \qquad (1.22)$$

where N is the number of partitions of the spectrum. Now the integral can be approximated as a sum,

$$\bar{\eta}(x,t) = \sum_{n=1}^N \cos(\bar{k}_n x - \bar{\omega}_n t + \varepsilon_n)\sqrt{A^2(\bar{\omega}_n)\Delta\omega_n}, \qquad (1.23)$$

where $\bar{\omega}_n^2 = \bar{k}_n g$ (deep water) and $\bar{()}$ represents the average of each parameter. Let $S(\omega_n)$ represent the cumulative area under the spectral density curve, or

$$S(\omega_n) = \sum_{n=1}^N A^2(\omega_n)\Delta\omega_n. \qquad (1.24)$$

Thus,

$$A^2(\bar{\omega}_n)\Delta\omega_n \approx S(\omega_n) - S(\omega_{n-1}) = a^2, \qquad (1.25)$$

where a^2 is a constant to be determined. It follows that

$$Na^2 = S(\omega_N) \approx S(\infty) = \int_0^\infty A^2(\omega)d\omega. \qquad (1.26)$$

The Pierson-Moskowitz spectrum for the wave height is of the form

$$S\eta(\omega) = A^2(\omega) = \frac{A_0}{\omega^5}e^{-B/\omega^4}, \qquad (1.27)$$

where A_0 and B are constants defined by

$$A_0 = 8.1 \times 10^{-3}g^2 \qquad (1.28)$$

$$B = \frac{3.11}{H_s^2}, \tag{1.29}$$

The significant wave height H_s is an unidentifiable wave form that propagates like a physical wave. It is the mean value of the highest one-third of the waves in the wave record. It corresponds well with visual observation. From equations (1.24) and (1.27),

$$S(\omega) = \frac{A_0}{4B}e^{-B/\omega^4}, \tag{1.30}$$

and, therefore,

$$S(\infty) = \frac{A_0}{4B} \tag{1.31}$$

$$a^2 = \frac{A_0}{4BN}. \tag{1.32}$$

Now the partition frequencies can be determined for $\omega = F$,

$$\frac{A_0}{4B} = e^{B/F^4}S(F). \tag{1.33}$$

Because the partition is of equal areas

$$S(\omega_n) = \frac{n}{N}S(F) = \frac{A_0}{4B}e^{-B/\omega_n^4} = S(F)e^{B/F^4}e^{-B/\omega_n^4}, \tag{1.34}$$

and it follows that

$$\frac{N}{n}e^{B/F^4} = e^{B/\omega_n^4}. \tag{1.35}$$

Solving equation (1.35) for ω_n leads to the partition frequencies

$$\omega_n = \left(\frac{B}{\ln(N/n) + B/F^4}\right)^{0.25}, \qquad n = 1, 2, ..., N. \tag{1.36}$$

Therefore, the wave elevation $\eta(x, t)$ can be approximated by

$$\bar{\eta}(x, t) = a \sum_{n=1}^{N} \cos(k_n x - \omega_n t + \varepsilon_n). \tag{1.37}$$

The wave loading on the tower is a function of wave velocity and acceleration. Therefore, in the numerical studies, these have to be expressed as functions of the approximate wave elevation. Thus, in the expressions for wave velocity and acceleration, the following substitutions are made

$$\left.\begin{array}{c} H \Longrightarrow \sqrt{\frac{A_0}{4BN}} \\ \omega \Longrightarrow \omega_n \\ k \Longrightarrow k_n, \end{array}\right\}$$

where $n = 1, 2, ..., N$, and H being the wave height sometimes denoted as a.

As mentioned earlier, the wave height is assumed to have the Pierson-Moskowitz spectrum, specifically

$$S\eta(\omega) = \frac{0.0081g^2}{\omega^5}\exp(-\frac{3.11}{H_s^2\omega^4}). \tag{1.38}$$

The significant wave height, H_s, changes the maximum wave height and the wave frequency distribution as shown in Fig. 1.2 for $H_s = 15$ and $9\ m$.

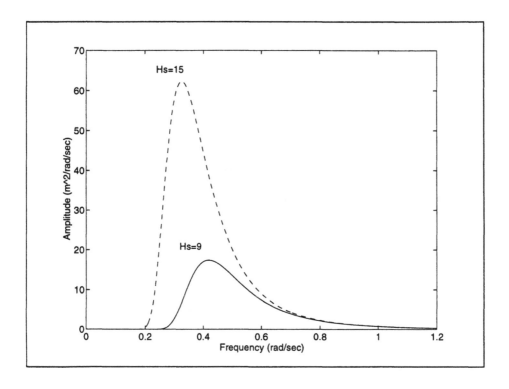

1.2 Example for Two Pierson-Moskowitz Wave Height Spectra.

Dynamics of the Structure

A generic structural model is chosen for the general formulation, one which forms the basis for further specification in the subsequent chapters. The notations that are used for derivatives are the subscript notation as well as the overdot for derivatives with respect to time.

Introduction

The schematic of the model considered is shown in Fig. 1.3. The model consists of a pipe conveying fluid that is submerged in the ocean. The pipe is connected to a buoyant deck having a transverse moment of inertia J_p and mass M. The pipe's lateral deflection is denoted $y(x, t)$. The pipe is subjected to velocities of wave $\mathbf{u}_w$, current $\mathbf{U}_c$, and wind $\mathbf{U}_w$. Base excitation is also included in the model via w, u displacements in the x, y directions, respectively. The general nonlinear differential equations of motion are derived using Hamilton's principle. All forces/moments, velocities and accelerations are derived and expressed in the fixed coordinate system x, y and are evaluated at the instantaneous position of the pipe, causing the equations of motion to be highly nonlinear. This model is sufficiently general to represent either the partially submerged articulated tower, a tension leg platform, a riser conveying fluid or an appendage connected to the structure's deck. This is done by changing the boundary conditions, the external forces and some other physical parameters. It will be shown, in later chapters, how this generic model is used for each application.

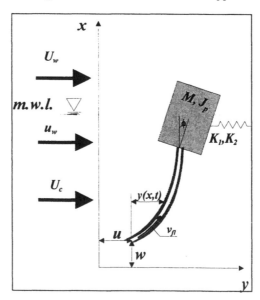

1.3 Schematic of the Model.

Hamilton's Principle

The equations of motion are derived under several assumptions;

- The pipe is inextensible and always in tension.
- Planar motion only.
- The pipe's length is much larger than its diameter.
- The water depth is greater than the pipe's length.
- The internal fluid velocity is fully developed but time dependent.
- The pipe is a slender and smooth structure with circular cross section.
- The pipe may be subjected to base and/or top excitation, wave, current and wind loads.
- The current and wind velocities propagate in the direction of the wave.

Since the model studied is continuous, Hamilton's principle is used to derive the equation of motion. Hamilton's principle, in the presence of external forces, states that (see *Lagrangian Dynamics*, pp. 333 [5])

$$\int_{t_1}^{t_2} [\delta\mathcal{L} + \Sigma \delta q_i F_{q_i}] \, dt = 0,$$

where the Lagrangian $\mathcal{L}$ is

$$\mathcal{L} = E_K - E_P, \tag{1.39}$$

and E_K and E_P are the kinetic and potential energies, respectively. F_{q_i} is the external force in the direction of the generalized coordinate q_i.

Derivation of the Lagrangian

To derive the Lagrangian $\mathcal{L}$, the potential and kinetic energies must be found. A schematic of the geometry of the pipe and the deck connected to it is shown in Fig. 1.4, where an element $d\bar{m}$ of the pipe undergoes lateral displacement $y(x, t)$ and rotation $\theta(x, t)$ about the z axis.

The following geometrical relations are derived from Fig. 1.5 and are used in this part of the derivation:

$$ds = \sqrt{dx^2 + dy^2} = dx\sqrt{1 + y_x^2} \tag{1.40}$$

$$\sin\theta = \frac{y_x}{\sqrt{1 + y_x^2}} \tag{1.41}$$

$$\cos\theta = \frac{1}{\sqrt{1 + y_x^2}} \tag{1.42}$$

$$\tan\theta = y_x. \tag{1.43}$$

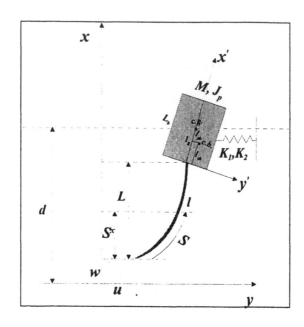

1.4 Schematic and Geometry.

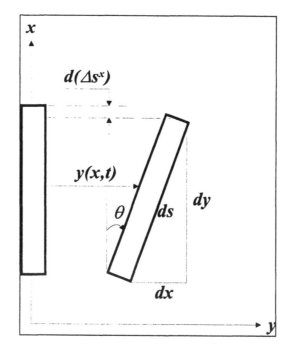

1.5 Schematic of an Element.

Fig. 1.6 shows a free body diagram, from which the reaction forces F_x^R, F_y^R, and M^R are derived. These forces are used to formulate the natural boundary conditions, as well as the general expression for the external work.

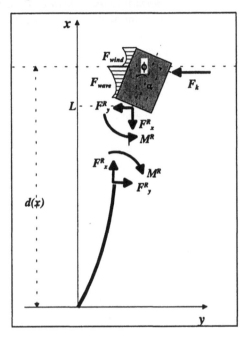

1.6 Free Body Diagram of the Inertia at the Top.

The deck is assumed rigid having three degrees of freedom about the point of attachment. Thus, three equations can be written, two force equations (for transverse motion) and one moment equation (for rotational motion):

$$M \left(L_{tt} - \frac{1}{2}\varphi_{tt}L_b \sin\varphi - \frac{1}{2}\varphi_t^2 L_b \cos\varphi \right) - (F_{wind} + F_{wave}) \sin\varphi =$$

$$Mg + F_x^R - F_b^D \tag{1.44}$$

$$(F_{wind} + F_{wave}) \cos\varphi - M \left(\dot{V}_T^y - \frac{1}{2}\varphi_{tt}L_b \sin\varphi - \frac{1}{2}\varphi_t^2 L_b \cos\varphi \right) =$$

$$F_k + F_y^R$$

$$\frac{1}{2}ML_b \left(L_{tt} \sin\varphi + L_t\varphi_t \cos\varphi + \dot{V}_T^y \sin\varphi + V_T^y\varphi_t \cos\varphi \right) - J_p\varphi_{tt} -$$

$$F_k \left(\alpha \cos\varphi - \frac{1}{2}D_b \sin\varphi \right) + M_{wind} + M_{wave} = M^R - M_{gb}, \tag{1.45}$$

where D_b is the diameter of the (buoyant) deck, L_b is the deck height, L is the projection of the pipe (in the x direction) and α is the perpendicular distance (in the x'

direction) between the point of attachment of the cable to the deck and the connection point of the spring, L_{tt} and $\dot{V}_T^y$ are the accelerations in the x and y directions, respectively, of the point of attachment. F_{wind} and F_{wave} are the wind and wave forces, respectively, per unit length acting *normal to the structure*. F_k is the spring force and F_x^R, F_y^R are the reaction forces at the point of attachment (see Fig. 1.6). M^R, M_{gb} are the reaction moment and the moment due to gravity and buoyancy. J_p is the moment of inertia of the deck with respect to the point of attachment and φ_{tt} is the angular acceleration about this point. M_{wind} and M_{wave} are the wind and the wave moments about the point of attachment. Explicit expressions for all terms are to be derived subsequently.

The time–dependent buoyancy force of the deck, F_b^D, is given by

$$F_b^D = \rho g \hat{V}_1, \tag{1.46}$$

where $\hat{V}_1$ is the volume of the submerged part of the deck,

$$\hat{V}_1 = \int_0^{L_s} A_B(x')dx', \tag{1.47}$$

where x' is the distance measured along the length of the circular cylindrical deck, and $A_B(x')$ is the area of the cross section of the deck. Thus, the buoyancy force due to the deck is

$$F_b^D = \int_0^{L_s} \rho g A_B(x')dx'. \tag{1.48}$$

L_s is the length of the submerged part of the deck in the x' direction,

$$L_s = \frac{d(x) - L}{\cos \varphi}, \tag{1.49}$$

and $d(x)$ is the water depth. From equation (1.44), the reaction forces in the x and y directions are found to be

$$F_x^R = \int_L^{\frac{d(x)-L}{\cos\varphi}} \rho g A_B(x')dx' - (F_{wind} + F_{wave})\sin\varphi - Mg$$

$$+ M(L_{tt} - \frac{1}{2}\varphi_{tt}L_b \sin\varphi - \frac{1}{2}\varphi_t^2 L_b \cos\varphi) \tag{1.50}$$

$$F_y^R = (F_{wind} + F_{wave})\cos\varphi - F_k - M(\dot{V}_T^y - \frac{1}{2}\varphi_{tt}L_b \cos\varphi - \frac{1}{2}\varphi_t^2 L_b \sin\varphi)$$

and the reaction moment from equation (1.45) is

$$M^R = M_{gb} + M_{wind} + M_{wave} - F_k(\alpha \cos\varphi - \frac{1}{2}D_b \sin\varphi) - J_p\varphi_{tt}$$

$$+ \frac{1}{2}ML_b(L_{tt}\sin\varphi + L_t\varphi_t \cos\varphi + \dot{V}_T^y \cos\varphi + V_T^y\varphi_t \sin\varphi). \tag{1.51}$$

The water depth $d(x)$ is given by

$$d(x) = d_0 + \eta(x,t), \tag{1.52}$$

where d_0 is the mean water depth and $\eta(x,t)$ is the wave elevation at the instantaneous position of the deck, given by

$$\eta(x,t) = \frac{1}{2}H\cos(ky(x,t) - \omega t), \tag{1.53}$$

where H is the significant wave height, ω being the wave frequency and k the wave number, which are related by

$$\omega^2 = gk\tan(kd_0). \tag{1.54}$$

The various energies and kinematics needed are derived next.

Bending Strain Energy: The general expression for the bending strain energy due to bending is given by

$$BE = \int_0^l \frac{M_b^2}{2EI(s)}ds, \tag{1.55}$$

where M_b is the bending moment, E is Young's modulus $I(s)$ is the pipe's cross section moment of inertia and s is a coordinate measured along the pipe. The bending moment is given by

$$M_b = EI(s)\frac{1}{r(s)}, \tag{1.56}$$

with $r(s)$ being the radius of curvature,

$$r(s) = \frac{ds}{d\theta}. \tag{1.57}$$

Transforming to Cartesian coordinates results in

$$\frac{1}{r(x)} = \frac{y_{xx}}{(1 + y_x^2)^{3/2}}. \tag{1.58}$$

Substituting equation (1.58) into (1.56) and that result into equation (1.55) leads to the expression for the strain energy in terms of the displacement $y(x,t)$

$$BE = \int_w^L \frac{1}{2}EI(x)\frac{y_{xx}^2}{(1 + y_x^2)^3}dx, \tag{1.59}$$

where L is the projection of the length of the pipe in the x direction such that the length of the pipe l is

$$l = \int_w^L \frac{1}{\cos\theta}dx = \int_w^L \sqrt{1 + y_x^2}dx. \tag{1.60}$$

Potential Energy: The potential energy due to gravity and buoyancy is given by

$$PE = \int_w^L F_{gb}(x,t)d(\Delta s^x), \tag{1.61}$$

where $F_{gb}(x,t)$ is the total force in the x direction due to gravity and buoyancy. $d(\Delta s^x)$ is the projection of an element ds onto x with respect to the rigid body configuration (see Fig. 1.5):

$$\Delta s^x = s - s^x = \int_w^{s^x} \sqrt{1 + y_x^2}dx - s^x \tag{1.62}$$

so that

$$\Delta s^x = \int_w^{s^x} \left(\sqrt{1 + y_x^2} - 1 \right) dx, \tag{1.63}$$

where s^x is the projection of a segment with arc length s onto the x axis (see Fig. 1.4). Therefore,

$$d(\Delta s^x) = \frac{\partial (\Delta s^x)}{\partial x} dx = \left(\sqrt{1 + y_x^2} - 1 \right) dx. \tag{1.64}$$

The gravity force acting on the pipe is

$$F_g(x) = - \left[\int_{s^x}^L \left(\rho_T A_T(x) + \rho_{fl} A_{fl}(x) \right) dx \right] (g + w_{tt}), \tag{1.65}$$

where ρ_T and ρ_{fl} are the densities of pipe or tube and the internal fluid, respectively. The area of the cross sections of the pipe or tube and the inner flow area are $A_T(x)$ and $A_{fl}(x)$, respectively,

$$A_T(x) = \frac{1}{4}\pi \left(D_o^2(x) - D_i^2(x) \right) \tag{1.66}$$

$$A_{fl}(x) = \frac{1}{4}\pi D_i^2(x). \tag{1.67}$$

The total area of the pipe is

$$A(x) = A_T(x) + A_{fl}(x) = \frac{1}{4}\pi D_o^2(x), \tag{1.68}$$

where $D_o(x)$ and $D_i(x)$ are the outer and inner diameters of the pipe.

The buoyancy force acting on the pipe is given by

$$F_b(x) = \rho g \hat{V}_2 - p A_{fl}(x), \tag{1.69}$$

where p is the internal pressure, ρ is the density of the external fluid and $\hat{V}_2$ is the volume of the submerged pipe, given by

$$\hat{V}_2 = \int_w^L \left(A_T(x) + A_{fl}(x) \right) dx, \tag{1.70}$$

so that the total buoyancy force is

$$F_b(x) = \rho g \int_{s^x}^L \left(A_T(x) + A_{fl}(x) \right) dx - p A_{fl}(x). \tag{1.71}$$

Adding the gravity and buoyancy forces results in

$$F_{gb}(x) = \int_{s^x}^L \left[\begin{array}{c} (A_T(x) + A_{fl}(x)) \rho g - \\ (\rho_T A_T(x) + \rho_{fl} A_{fl}(x)) (g + w_{tt}) \end{array} \right] dx$$
$$+ \int_0^{\frac{d(x) - L}{\cos \varphi}} \rho g A_B(x') dx' - Mg - p A_{fl}(x). \tag{1.72}$$

Finally, substituting equations (1.72) and (1.64) into (1.61) leads to the expression for the potential energy

$$PE = \int_w^L \left\{ \int_{s^x}^L \left[\begin{array}{c} (A_T(x) + A_{fl}(x))\, \rho g - \\ (\rho_T A_T(x) + \rho_{fl} A_{fl}(x))\, (g + w_{tt}) \end{array} \right] dx \right. \tag{1.73}$$

$$\left. + \int_0^{\frac{d(x)-L}{\cos \varphi}} \rho g A_B(x') dx' - Mg - pA_{fl}(x) \right\} \left(\sqrt{1 + y_x^2} - 1 \right) dx.$$

Kinetic Energy: The kinetic energy consists of two components, one due to rectilinear velocity and the other due to angular velocity. It is derived by following a pipe or tether element $d\bar{m}_T$ and an internal fluid element $d\bar{m}_{fl}$ from the non–perturbed state to the perturbed one. Therefore the continuity equation is identically satisfied. The pipe's kinetic energy due to rectilinear velocity is

$$KE_T^L = \int_w^L \frac{1}{2} d\bar{m}_T \left[(V_T^x)^2 + (V_T^y)^2 \right], \tag{1.74}$$

where $d\bar{m}_T$ is the mass per unit length of an element in the perturbed system, including the added mass term due to the fluid,

$$d\bar{m}_T = [\rho_T A_T(x) + C_A \rho A(x)]\, dx, \tag{1.75}$$

and C_A is the added mass coefficient.

V_T^x and V_T^y are the absolute velocities of the pipe's element dx in the x and y directions. Since the pipe's base is not stationary, its motion can actually be considered as a moving continua problem. For a general function $G(x, t)$ the material derivatives with respect to t are defined as follows (see Bar–Avi and Porat [6]),

$$\frac{DG}{Dt} = G_t + \frac{dx}{dt} G_x \tag{1.76}$$

$$\frac{D^2 G}{Dt^2} = G_{tt} + 2\frac{dx}{dt} G_{xt} + \left(\frac{dx}{dt} \right)^2 G_{xx} + G_x \left(\frac{d^2 x}{dt^2} + \frac{d^2 x}{dxdt} \right), \tag{1.77}$$

where dx/dt is the velocity of the reference point in position x of the field. Thus the absolute velocities in the x and y direction are obtained,

$$V_T^x = s_t^x + w_t(1 + s_x^x) \tag{1.78}$$

$$V_T^y = y_t + w_t y_x + u_t, \tag{1.79}$$

where u_t and w_t are the base velocities. Substituting equations (1.78) and (1.79) into equation (1.74) results in the expression for the rectilinear kinetic energy expressed in the inertial coordinate system x, y,

$$KE_T^L = \int_w^L \frac{1}{2} [\rho_T A_T(x) + C_A \rho A(x)] \cdot \tag{1.80}$$

$$\left\{ (s_t^x + w_t(1 + s_x^x))^2 + (y_t + w_t y_x + u_t)^2 \right\} dx.$$

The angular velocity term is

$$KE_T^A = \int_w^L \frac{1}{2} dI_T(x) \theta_t^2, \tag{1.81}$$

where $dI_T(x)$ is the mass moment of inertia of the cross section of the element $d\bar{m}_T$, defined as (see Timoshenko and Young pp. 431-433 [7])

$$dI_T(x) = d\bar{m}_T \frac{I_T(x)}{A_T(x)} = [\rho_T A_T(x) + C_A \rho A(x)] \frac{\frac{\pi}{64}(D_o^4(x) - D_i^4(x))}{\frac{\pi}{4}(D_o^2(x) - D_i^2(x))} dx$$

$$= \frac{1}{4\pi} [\rho_T A_T(x) + C_A \rho A(x)] (A(x) + A_{fl}(x)) dx. \qquad (1.82)$$

The angular velocity θ_t is calculated as follows. Starting with equations (1.41), (1.42), (1.43),

$$\tan \theta(x,t) = y_x(x,t), \qquad (1.83)$$

and taking the partial time derivative of both sides

$$\theta_t (1 + \tan^2 \theta) = y_{xt}, \qquad (1.84)$$

leads to the angular velocity of the element $d\bar{m}_T$

$$\theta_t = \frac{y_{xt}}{1 + y_x^2}. \qquad (1.85)$$

Substituting equations (1.82), (1.85) into (1.81) results in the expression for the angular kinetic energy of the pipe in the inertial coordinate system x, y,

$$KE_T^A = \int_w^L \frac{1}{2} \frac{1}{4\pi} [\rho_T A_T(x) + C_A \rho A(x)] (A(x) + A_{fl}(x)) \frac{y_{xt}^2}{(1 + y_x^2)^2} dx. \qquad (1.86)$$

Thus, the total kinetic energy of the pipe is

$$KE_T = \frac{1}{2} \{ \int_w^L [\rho_T A_T(x) + C_A \rho A(x)] \cdot \qquad (1.87)$$

$$\left[(s_t^x + w_t(1 + s_x^x))^2 + (y_t + w_t y_x + u_t)^2 \right]$$

$$+ \frac{1}{4\pi} [\rho_T A_T(x) + C_A \rho A(x)] (A(x) + A_{fl}(x)) \frac{y_{xt}^2}{(1 + y_x^2)^2} \} dx.$$

The kinetic energy of the internal fluid due to rectilinear velocity is

$$KE_{fl}^L = \int_w^L \frac{1}{2} d\bar{m}_{fl} \left[(V_{fl}^x)^2 + (V_{fl}^y)^2 \right], \qquad (1.88)$$

where $d\bar{m}_{fl}$ is the mass of the fluid element,

$$d\bar{m}_{fl} = \rho_{fl} A_{fl}(x) dx. \qquad (1.89)$$

The rectilinear velocities in the x and y directions are obtained by adding the components of pipe velocity to the components of fluid velocity

$$V_{fl}^x = V_T^x + v_{fl} \frac{1}{\sqrt{1 + y_x^2}} \qquad (1.90)$$

$$V_{fl}^y = V_T^y + v_{fl} \frac{y_x}{\sqrt{1 + y_x^2}}. \qquad (1.91)$$

Substituting the pipe's velocities (equations (1.78), (1.79)) into equations (1.90) and (1.91) results in the fluid's rectilinear velocities,

$$V_{fl}^x = s_t^x + w_t(1 + s_x^x) + v_{fl}\frac{1}{\sqrt{1 + y_x^2}} \qquad (1.92)$$

$$V_{fl}^y = y_t + \left[w_t + v_{fl}\frac{1}{\sqrt{1 + y_x^2}}\right]y_x + u_t. \qquad (1.93)$$

Thus, the rectilinear kinetic energy is found by substituting equations (1.92), (1.93) and (1.89) into (1.88),

$$KE_{fl}^L = \frac{1}{2}\int_w^L \{\rho_{fl}A_{fl}(x)[s_t^x + w_t(1 + s_x^x) + v_{fl}\frac{1}{\sqrt{1 + y_x^2}}]^2 \qquad (1.94)$$

$$+[y_t + (w_t + v_{fl}\frac{1}{\sqrt{1 + y_x^2}})y_x + u_t]^2\}dx.$$

The angular velocity term for the fluid is derived in the same way as the one for the pipe:

$$KE_T^A = \int_w^L \frac{1}{2}dI_{fl}(x)\theta_t^2, \qquad (1.95)$$

where dI_{fl} is given by

$$dI_{fl}(x) = d\bar{m}_{fl}\frac{I_{fl}(x)}{A_{fl}(x)} = \frac{1}{4\pi}\rho_{fl}A_{fl}^2(x)dx. \qquad (1.96)$$

Therefore, the angular kinetic energy is

$$KE_{fl}^A = \int_w^L \frac{1}{2}\frac{1}{4\pi}\rho_{fl}A_{fl}^2(x)\frac{y_{xt}^2}{(1 + y_x^2)^2}dx, \qquad (1.97)$$

and the total kinetic energy of the internal fluid is

$$KE_{fl} = \frac{1}{2}\int_w^L \left\{\rho_{fl}A_{fl}(x)\left[(s_t^x + w_t(1 + s_x^x) + v_{fl}\frac{1}{\sqrt{1 + y_x^2}})^2 \right.\right. \qquad (1.98)$$

$$\left.\left. +(y_t + (w_t + v_{fl}\frac{1}{\sqrt{1 + y_x^2}})y_x + u_t)^2\right] + \frac{1}{4\pi}\rho_{fl}A_{fl}^2(x)\frac{y_{xt}^2}{(1 + y_x^2)^2}\right\}dx.$$

The Lagrangian: Substituting the expressions for the energies into equation (1.39), the Lagrangian is finally found to be given by the equation

$$\mathcal{L} = \frac{1}{2} \int_w^L \{ [\rho_T A_T(x) + C_A \rho A(x)][(s_t^x + w_t(1 + s_x^x))^2 + (y_t + w_t y_x + u_t)^2]$$

$$+ \frac{1}{4\pi} [\rho_T A_T(x) + C_A \rho A(x)] \, (A(x) + A_{fl}(x)) \frac{y_{xt}^2}{(1 + y_x^2)^2} +$$

$$\frac{1}{4\pi} A_{fl}^2(x) \frac{y_{xt}^2}{(1 + y_x^2)^2} + \rho_{fl} A_{fl}(x) \left[s_t^x + w_t(1 + s_x^x) + v_{fl} \frac{1}{\sqrt{1 + y_x^2}} \right]^2 +$$

$$\rho_{fl} A_{fl}(x) \left[y_t + (w_t + v_{fl} \frac{1}{\sqrt{1 + y_x^2}}) y_x + u_t \right]^2 - EI(x) \frac{y_{xx}^2}{(1 + y_x^2)^3} -$$

$$\{ \int_{s^x}^L [(A_T(x) + A_{fl}(x))\rho g - (\rho_T A_T(x) + \rho_{fl} A_{fl}(x))(g + w_{tt})] \, dx +$$

$$\int_0^{\frac{d(x)-L}{\cos \varphi}} \rho g A_B(x') dx' - M g - p A_{fl}(x) \} (\sqrt{1 + y_x^2} - 1) \} dx, \qquad (1.99)$$

where all terms have been previously defined. The fully nonlinear partial differential equation of motion for the lateral displacement, $y(x,t)$, of the pipe can be formulated by taking the variation of the Lagrangian, $\delta \mathcal{L}$. However, a rather simplified equation is formulated later in this chapter. In general, together with the equation for the lateral deflection, three more equations of motion for the deck are needed. Two for the transverse motion, heave and surge, from equations (1.50), and the third for the rotational motion, from equation (1.51). In this analysis it is assumed that the pipe is inextensible and always in tension. These assumptions constrain the transverse motion of the deck to the lateral motion of the pipe. Therefore, the two transverse equations of motion for the deck (equation (1.50)) are incorporated into the boundary conditions, and only the rotation equation for the deck is needed.

Boundary Conditions

The above formulation is general enough to allow different boundary conditions. Both the bottom and the top ends are either pinned or fixed. The bottom to the sea floor and the top to the deck. The mathematical form of the boundary conditions is:

a.   Pinned,

$$x = L + w; \quad \begin{cases} S + F_y^R \cos \theta - F_x^R \sin \theta = 0 \\ M_b = M^R = 0 \end{cases} \qquad (1.100)$$

$$x = w; \quad \begin{cases} y = 0 \\ M_b = M^R = 0 \end{cases} \qquad (1.101)$$

b.   Fixed

$$x = L + w; \quad \begin{cases} M_b + M^R = 0 \\ S + F_y^R \cos \theta - F_x^R \sin \theta = 0 \end{cases} \qquad (1.102)$$

$$x = w; \quad \begin{cases} y = 0 \\ y_x = 0 \end{cases} \qquad (1.103)$$

As a reminder, M_b is the pipe bending moment, M^R is the reaction moment (see equation (1.51)), S the shear force, and D_b is the diameter of the deck. The forces and moments used in the boundary conditions, equations (1.100), (1.101), (1.102) and (1.103), S, F_y^R, F_x^R, M_b and M^R,

are functions of the deck kinematics and dynamics, that is, $\dot{V}_T^y$, F_k, F_{wave}^D, F_{wind}^D and others. These terms are derived next.

The spring force on the deck is assumed to be bilinear, meaning that the stiffness is different for positive and negative displacements (see Fig. (1.4)),

$$F_k = \begin{cases} k_1[y(L) + \alpha \sin\varphi - \frac{D_b}{2}(1 - \cos\varphi)] & \dot{y} > 0 \\ k_2[y(L) + \alpha \sin\varphi - \frac{D_b}{2}(1 - \cos\varphi)] & \dot{y} < 0, \end{cases} \tag{1.104}$$

where k_1 and k_2 are the different stiffnesses. The wind and wave forces on the deck are

$$F_{wave}^D = \int_0^{L_s} \hat{F}_{wave}^D dx' \tag{1.105}$$

$$F_{wind}^D = \int_{d(x)}^{L_b - L_s} \hat{F}_{wind}^D dx', \tag{1.106}$$

where the wind force per unit length of deck, $\hat{F}_{wind}^D$, is given by

$$\hat{F}_{wind}^D = \frac{1}{2} C_{D_a} \rho_a D_b |(U_w^n - V_D^n)|(U_w^n - V_D^n), \tag{1.107}$$

where ρ_a is the air density, C_{D_a} is the air drag coefficient and U_w^n and V_D^n are the wind and deck velocities in a direction normal to the deck,

$$U_w^n = U_w \cos\varphi \tag{1.108}$$

$$V_D^n = (y_t(L) + u_t) \cos\varphi + x' \varphi_t - (w_t + L_t) \sin\varphi. \tag{1.109}$$

Substituting equation (1.108) and equation (1.109) into equation (1.107) yields

$$\hat{F}_{wind}^D = \frac{1}{2} C_{D_a} \rho_a D_b \{ |((U_w - y_t(L) - u_t) \cos\varphi - x' \varphi_t - (w_t + L_t) \sin\varphi)| \cdot$$
$$((U_w - y_t(L) - u_t) \cos\varphi - x' \varphi_t - (w_t + L_t) \sin\varphi) \}, \tag{1.110}$$

and integrating along x' in equation (1.106) results in the total wind force,

$$F_{wind}^D = \frac{1}{6\varphi_t} \left\{ \frac{[\cos^2\varphi(-U_w + y_t(L) + u_t) + \varphi_t L_b \cos\varphi - \varphi_t d + \varphi_t L]^3}{\cos^3\varphi} \right.$$
$$\left. - [\cos\varphi(-U_w + y_t(L) + u_t) + \varphi_t d]^3 \right\} C_{D_a} \rho_a D_b. \tag{1.111}$$

The wave force per unit length of deck, $\hat{F}_{wave}^D$, is

$$\hat{F}_{wave}^D = \frac{1}{2} C_D \rho D_b |V_{fl}^n - V_D^n|(V_{fl}^n - V_D^n) + \frac{1}{4} C_M \rho \pi D_b^2 \dot{V}_{fl}^n, \tag{1.112}$$

where V_{fl}^n and $\dot{V}_{fl}^n$ are the wave velocity and acceleration normal to the deck, given by

$$V_{fl}^n = u_{wave} \cos\varphi - w_{wave} \sin\varphi \tag{1.113}$$

$$\dot{V}_{fl}^{n} = \dot{u}_{wave}\cos\varphi - w_{twave}\sin\varphi. \tag{1.114}$$

The wave force is evaluated by substituting equations (1.113) and (1.114) into equation (1.105) and integrating. Thus the resulting moments are

$$M_{wave}^{D} = \int_{0}^{L_{s}} F_{wave}^{D} x^{'} dx^{'} \tag{1.115}$$

$$M_{wind}^{D} = \int_{d}^{L_{b}-L_{s}} F_{wind}^{D} x^{'} dx^{'} \tag{1.116}$$

$$M_{gb}^{D} = \int_{0}^{L_{s}} \rho g A_{B}(x^{'}) dx^{'} l_{b} - M g l_{g}. \tag{1.117}$$

The moment arm of the buoyancy force, l_b, is

$$l_{b} = l_{xb}\sin\varphi + l_{yb}\cos\varphi, \tag{1.118}$$

and for a cylindrical cross section (see Fig. 1.4),

$$l_{yb} = \frac{D_{o}^{2}}{16L_{s}}\tan\varphi \tag{1.119}$$

$$l_{xb} = \frac{1}{2}L_{s} + \frac{D_{o}^{2}}{32L_{s}}\tan^{2}\varphi. \tag{1.120}$$

Therefore, substituting equations (1.49), (1.119) and (1.120) into equation (1.118) results in

$$l_{b} = \frac{1}{2}\tan\varphi\frac{(d(x) - L)^{2} + D_{o}^{2}\left(1 + \cos^{2}\varphi\right)}{d(x) - L}. \tag{1.121}$$

The gravity moment arm is

$$l_{g} = \frac{1}{2}L_{b}\sin\varphi. \tag{1.122}$$

The pipe's lateral acceleration $\dot{V}_T^y$ is obtained by total differentiation of the lateral velocity, V_T^y, equation (1.79) with respect to time,

$$\begin{aligned}\dot{V}_{T}^{y} &= \frac{\partial V_{T}^{y}}{\partial t} + \frac{dx}{dt}\frac{\partial(V_{T}^{y})}{\partial x} \\ &= y_{tt} + 2w_{t}y_{xt} + w_{t}^{2}y_{xx} + w_{tt}y_{x} + u_{tt}.\end{aligned} \tag{1.123}$$

The shear force S and the normal force N are determined from Fig. 1.7, which depicts the forces and moments acting on an element ds of the pipe.

The element of pipe is in equilibrium, so in order to find the relation between the shear force S and the displacement y and its derivative, the moment equation about the centroid of the element is formed,

$$\sum M_{0} = I_{T}(x)dx\theta_{tt}, \tag{1.124}$$

where $I_T(x)dx\theta_{tt}$ is the inertial moment due to the angular acceleration of the element, and $\sum M_0$ is the sum of the external moments about the center of the element.

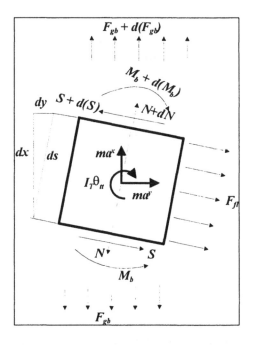

1.7 Forces and Moments Acting on an Element.

The angular acceleration, θ_{tt}, is derived by taking the total time derivative of the angular velocity θ_t given in equation (1.85),

$$
\begin{aligned}
\theta_{tt} &= \frac{\partial}{\partial t}\left(\frac{y_{xt}}{1+y_x^2}\right) + w_t \frac{\partial}{\partial x}\left(\frac{y_{xt}}{1+y_x^2}\right) \\
&= \frac{y_{xtt}}{1+y_x^2} - \frac{2\,y_{xt}^2\,y_x}{(1+y_x^2)^2} + w_t\left(\frac{y_{xxt}}{1+y_x^2} - \frac{2\,y_{xt}\,y_x y_{xx}}{(1+y_x^2)^2}\right).
\end{aligned} \tag{1.125}
$$

and the sum of moments is

$$
\begin{aligned}
\sum M_0 &= M_b + dM_b - \frac{1}{2}(S+dS)ds - \frac{1}{2}(F_{gb}+dF_{gb})dy - M_b - \frac{1}{2}Sds \\
-\frac{1}{2}F_{gb}dy &= \frac{\partial M_b}{\partial x}dx - Sdx\sqrt{1+y_x^2} - F_{gb}y_x dx - \\
\frac{1}{2}\frac{\partial S}{\partial x}&(dx)^2\sqrt{1+y_x^2} - \frac{1}{2}\frac{\partial F_{gb}}{\partial x}y_x\,(dx)^2.
\end{aligned} \tag{1.126}
$$

Equating (1.124) into (1.126), and neglecting terms of order $(dx)^2$, results in an expression for the shear force,

$$
S = \left(\frac{\partial M_b}{\partial x} - F_{gb}y_x\right)\frac{1}{\sqrt{1+y_x^2}} - I_T(x)\theta_{tt}, \tag{1.127}
$$

where use has been made of the relation $ds = dx\sqrt{1+y_x^2}$, and the bending moment M_b is determined by substituting equation (1.58) into (1.56),

$$M_b = EI(x)\frac{y_{xx}}{(1+y_x^2)^{3/2}}. \tag{1.128}$$

Differentiating the bending moment with respect to x, we find

$$\frac{\partial M_b}{\partial x} = E[\frac{4I_x(x)y_{xx}}{(1+y_x^2)^{3/2}} + \frac{I(x)y_{xxx}}{(1+y_x^2)^{3/2}} - \frac{3I(x)y_{xx}^2 y_x}{(1+y_x^2)^{5/2}}]. \tag{1.129}$$

Substituting equations (1.125) and (1.129) with the expression for $I_T(x)$ into equation (1.127) results in the expression for the shear force,

$$S = E[\frac{4I_x(x)y_{xx}}{(1+y_x^2)^{3/2}} + \frac{I(x)y_{xxx}}{(1+y_x^2)^{3/2}} - \frac{3I(x)y_{xx}^2 y_x}{(1+y_x^2)^{5/2}}] - F_{gb}(x)\frac{y_x}{\sqrt{1+y_x^2}}$$
$$-I_T(x)[\frac{y_{xtt}}{1+y_x^2} - \frac{2y_{xt}^2 y_x}{(1+y_x^2)^2} + w_t(\frac{y_{xxt}}{1+y_x^2} - \frac{2y_{xt}y_xy_{xx}}{(1+y_x^2)^2})]. \tag{1.130}$$

Taking the Taylor expansion of S about $y = 0$ while retaining only linear terms, and assuming constant cross section, simplifies the expression for the shear force to the known one in linear elasticity,

$$S^s = EIy_{xxx} - F_{gb}y_x - I_T(x)(y_{xtt} + w_ty_{xxt}). \tag{1.131}$$

The first two terms are due to elastic strain and internal tension, and the third term is due to the rotational motion. Thus all the terms needed for the boundary conditions are known.

Using the expressions for the bending moment and the shear force along the pipe and the small deflection assumption, the axial stress and the shear stress can be evaluated. The axial stress is due to the bending moment and axial force is

$$\sigma = \frac{32M_b}{\pi D^3} - \frac{4F_{gb}}{\pi D^2}, \tag{1.132}$$

and the shear stress is

$$\tau = \frac{4S}{\pi D^2}, \tag{1.133}$$

where the expressions for M_b, F_{gb} and S were derived earlier. In order to perform a fatigue or reliability analysis, an equivalent stress is calculated using the Von-Mises formula,

$$\begin{aligned}\sigma_T &= \sqrt{\sigma^2 + 3\tau^2} \\ &= \sqrt{\left(\frac{32M_b}{\pi D^3} - \frac{4F_{gb}}{\pi D^2}\right)^2 + 3\left(\frac{4S}{\pi D^2}\right)^2}.\end{aligned} \tag{1.134}$$

More elaborate studies using the fully nonlinear equations are, of course, possible using numerical methods. Here the intension is to derive as many analytical expressions for key results as possible, recognizing that these are simplified.

Generalized Forces

The generalized forces are determined using the virtual work concept. Three types of forces are considered in this analysis, all acting in a direction normal to the pipe. The first acts on the pipe and the submerged part of the deck and is due to ocean current and gravity waves F_{fl}. The second is at the water level and is due to wave slamming F_s, and the third acts on the exposed part of the deck and is due to wind load F_w.

The generalized force due to an arbitrary external loading is found from the general relation

$$Q\delta y = F^n \delta n, \tag{1.135}$$

where F^n is any force acting normal to the pipe, and δn is the pipe's displacement in the direction of the force normal to the pipe, given by

$$\delta n = \delta x \frac{y_x}{\sqrt{1 + y_x^2}}. \tag{1.136}$$

Thus,

$$Q\delta y \doteq F^n \delta x \frac{y_x}{\sqrt{1 + y_x^2}}. \tag{1.137}$$

Dividing equation (1.137) by δx leads to,

$$Q\frac{\delta y}{\delta x} = Qy_x = F^n \frac{y_x}{\sqrt{1 + y_x^2}}, \tag{1.138}$$

and thus the generalized force Q is

$$Q = F^n \frac{1}{\sqrt{1 + y_x^2}}. \tag{1.139}$$

In the following subsections, we apply this general derivation to the various forces F^n that are needed to complete Lagrange's equation.

Wave Forces: The wave force per unit length due to drag and inertia is approximated via Morison's equation (see [2]). The added mass was included in the expression for the kinetic energy, equation (1.75),

$$F_{fl} = \frac{1}{2} C_D \rho D_o(x) \left| V_{fl}^n - V_T^n \right| (V_{fl}^n - V_T^n) + \frac{1}{4} C_M \rho \pi D_o^2(x) \dot{V}_{fl}^n, \tag{1.140}$$

where V_{fl}^n and $\dot{V}_{fl}^n$ are the absolute wave velocity and acceleration normal to the pipe, respectively, and V_T^n is the pipe's velocity in a direction normal to it,

$$V_{fl}^n = (u_{wave} + U_c)\cos\theta - w_{wave}\sin\theta \tag{1.141}$$

$$\dot{V}_{fl}^n = \dot{u}_{wave}\cos\theta - w_{twave}\sin\theta \tag{1.142}$$

$$V_T^n = V^y \cos\theta - V^x \sin\theta, \tag{1.143}$$

where w_{wave} and u_{wave} are the wave velocities in the x and y directions, and U_c is the current speed assumed to propagate in the y direction.

Substituting the trigonometric definitions equations (1.41), (1.42), (1.43) into equations (1.141)–(1.143) yields

$$V_{fl}^n = (u_{wave} + U_c - w_{wave}y_x)\frac{1}{\sqrt{1 + y_x^2}} \tag{1.144}$$

$$\dot{V}_{fl}^n = (\dot{u}_{wave} - w_{twave} y_x) \frac{1}{\sqrt{1 + y_x^2}} \tag{1.145}$$

$$V_T^n = (y_t + u_t) \frac{1}{\sqrt{1 + y_x^2}}. \tag{1.146}$$

Substituting equations (1.144)–(1.146) into equation (1.140) and the result in the form of equation (1.139) leads to the generalized gravity wave force per unit length. The generalized drag force, Q_D from Morison's equation, is then

$$Q_D = \frac{1}{2} C_D \rho D_o(x) \{ u_{wave} + U_c - w_{wave} y_x - y_t - u_t \} \cdot$$

$$\{ |u_{wave} + U_c - w_{wave} y_x - y_t - u_t| \} \frac{1}{(1 + y_x^2)^{3/2}}, \tag{1.147}$$

and substituting the 'deep water' wave velocities yields

$$Q_D = C_D \, \rho D_o(x) \frac{1}{2 \, (1 + y_x{}^2)^{3/2}} \cdot \tag{1.148}$$

$$\left| \frac{1}{2} H \omega e^{k(x-d)} \left(\cos(-ky + \omega t) + y_x \sin(-ky + \omega t) \right) + U_c - y_t - u_t \right| \cdot$$

$$\left(\frac{1}{2} H \omega e^{k(x-d)} \left(\cos(-ky + \omega t) + y_x \sin(-ky + \omega t) \right) + U_c - y_t - u_t \right).$$

Similarly, the inertia force, Q_M, is

$$Q_M = \frac{1}{4} C_M \pi \rho D_o^2(x) \left(\dot{u}_{wave} - w_{twave} y_x \right) \frac{1}{1 + y_x^2}, \tag{1.149}$$

and substituting the 'deep water' wave acceleration into equation (1.149) results in

$$Q_M = C_M \pi \, \rho D_o^2(x) \frac{1}{8} H \omega^2 \frac{e^{k(x-d)}}{(1 + y_x{}^2)} \left[-\sin(-ky + \omega t) + y_x \cos(-ky + \omega t) \right]. \tag{1.150}$$

Wind Force: The wind force is estimated in a similar way as the drag load except that the constants are different; see Faltinsen [8] . The wind drag force is given by

$$F_w = \frac{1}{2} C_{D_a} \rho_a D_b(x) |U_w^n - V_T^n| (U_w^n - V_T^n), \tag{1.151}$$

where U_w^n is the wind velocity normal to the pipe,

$$U_w^n = U_w \frac{1}{\sqrt{1 + y_x^2}}. \tag{1.152}$$

Substituting equation (1.152) into equation (1.151), and the result in the form of equation (1.139) leads to the expression for the wind generalized force,

$$Q_w = \frac{1}{2} C_{D_a} \rho_a D_b(x) |U_w - y_t - u_t| (U_w - y_t - u_t) \frac{1}{(1 + y_x^2)^{3/2}}. \tag{1.153}$$

The wind speed U_w is in the y direction and it is assumed to vary with height above mean water level. U_w is also assumed to be composed of two components, one deterministic and the other random. The deterministic velocity is related to the constant wind speed at $10\ m$, denoted by U_{10}, and the random velocity is related to the mean gust velocity at $10\ m$, that is $(U_{10})_{Gust}$, as follows (see Patel [9] pp. 186),

$$U_w = U_{10}\left(\frac{x}{10}\right)^{0.113} + (U_{10})_{Gust}\left(\frac{x}{10}\right)^{0.100}. \tag{1.154}$$

Wave Slamming Force: The wave slamming force is a periodic impulsive force. It has the same frequency as the wave frequency and a short duration of about $0.01\ s$. This force is formulated in a similar way as the drag force, with the wave's and pipe's absolute velocities evaluated along the impact length which is assumed to take place from the mean water level d to half of the wave height $\frac{1}{2}H$ (see Chakrabarti [10]). Therefore,

$$F_s = C_S\rho\frac{D_b(x)}{2}\ |\ V_{fl}^n - V_T^n\ |\ (V_{fl}^n - V_T^n), \tag{1.155}$$

where C_S is the wave slamming coefficient. Substituting the wave and pipe velocities into equation (1.155) and the result placed in the form of equation (1.139) leads to the wave slamming generalized force Q_s,

$$
\begin{aligned}
Q_s \ = \ & C_S\rho\frac{D_b(x)}{2}\left\{\left|\frac{1}{2}Hwe^{k(x-d)}(1-y_x) - y_t - u_t\right|\right\}\cdot \\
& \left\{\frac{1}{2}Hwe^{k(x-d)}(1-y_x) - y_t - u_t\right\}\frac{1}{(1+y_x^2)^{3/2}}.
\end{aligned}
\tag{1.156}
$$

Dissipative Force: The dissipative force is assumed to be viscous structural damping, thus

$$Q_{dis} = C_v y_t, \tag{1.157}$$

where C_v is the damping coefficient. The velocity used here is due to the pipe deflection only, since the structural damping is internal, and hence is assumed to be unaffected by the base excitation.

Governing Equations of Motion

The governing equations of motion are derived by setting the dynamic forces equal to the sum of external force defined by Q,

$$F_{dyn} = Q. \tag{1.158}$$

Both Q and F_{dyn} depend on the model being analyzed.

The generalized external forces Q equals the sum of the forces due to gravity waves, wave slamming, wind and viscous damping,

$$Q = Q_D + Q_M + Q_s + Q_w - Q_{dis}. \tag{1.159}$$

The expression for the dynamic and generalized forces is determined using 'MAPLE'.

To simplify the Lagrangian, moderate deformations and constant cross section are assumed. Using the Taylor expansion,

$$s = \int_w^{s^x} \sqrt{1 + y_x^2} dx \cong \int_w^{s^x} \left(1 + \frac{1}{2}y_x^2 + O[y_x]^4\right) dx. \tag{1.160}$$

Integrating by parts,

$$s \cong x - w + \frac{1}{2}\left(y(x)y_x(x) - y(w)y_x(w)\right) - \int_w^x y_x y_{xx} dx. \tag{1.161}$$

The integral is very small compared to the other terms, and the displacement $y(w) = 0$. Therefore,

$$s \cong x - w + \frac{1}{2}y(x)y_x(x) \tag{1.162}$$

$$l \cong L - w + \frac{1}{2}y(L)y_x(L) - x_0, \tag{1.163}$$

It can be seen that when $x = L \implies s = l$. x_0 is the heave motion of the pipe with zero lateral deflection. In this analysis it is assumed that x_0 is negligible, that is, the heave motion is caused only by lateral motion. With this assumption equation (1.50) is not necessary, thus only two equations of motion are needed, one for the lateral displacement and the other for the rotation of the deck. The expressions for L_t and L_{tt} are found by taking the time derivatives of L,

$$L_t = w_t - \frac{1}{2}\left(y_t(L)y_x(L) + y(L)y_{xt}(L)\right) \tag{1.164}$$

$$L_{tt} = w_{tt} - \frac{1}{2}\left(y_{tt}(L)y_x(L) + 2y_t(L)y_{xt}(L) + y(L)y_{xtt}(L)\right) \tag{1.165}$$

Since the fully nonlinear expressions are very complex, the dynamic forces are simplified using a Taylor expansion, neglecting third order and higher terms of the displacement, its derivatives with respect to x and/or t, and products of these derivatives, that is, y, y_x, y_{xx}, y_{xxx}, y_{xxxx}, y_{xt}, and y_{xxt} :

$$\begin{aligned}
&[A_T(\rho_T + C_A\rho) + \rho_{fl}A_{fl}](y_{tt} + u_{tt}) + EIy_{xxxx} + \\
&[2\rho_{fl}A_{fl}(w_t + v_{fl}) + A_T(\rho_T + C_A\rho)w_t]y_{xt} \\
&-[\rho_{fl}A_{fl}(w_{tt} + a_{fl}) + A_TC_A\rho w_{tt} + g(A_{fl}(\rho - \rho_{fl}) + A_T(\rho - \rho_T))]y_x \\
&+I_Ty_{xxtt} + [pA_{fl} - \rho g((A_{fl} + A_T)(l - x - w) \\
&+\frac{1}{\cos(\varphi)}A_b(d - L)) + \frac{3}{2}\rho_{fl}A_{fl}(2w_t + v_{fl})w_t \\
&+(M + (\rho_TA_T + \rho_{fl}A_{fl} + C_A\rho)(w + l - x)(g + w_{tt}) \\
&-A_T(\rho_T + C_A\rho)w_t^2]y_{xx} = Q,
\end{aligned} \tag{1.166}$$

where I_T is the mass moment of inertia and I is the moment of inertia of the cross section. The equation for the deck results by substituting equations (1.164) and (1.165) into equation (1.51),

$$-\frac{1}{2}ML_b[\left(w_{tt} - \frac{1}{2}\left(y_{tt}(L)y_x(L) + 2y_t(L)y_{xt}(L) + y(L)y_{xtt}(L)\right)\right)\sin\varphi$$

$$+\left(w_t - \frac{1}{2}\left(y_t(L)y_x(L) + y(L)y_{xt}(L)\right)\right)\varphi_t\cos\varphi$$

$$+\dot{V}_T^y\cos\varphi + V_T^y\varphi_t\sin\varphi] + J_p\varphi_{tt} = M_{gb} + M_{wind} + M_{wave}. \qquad (1.167)$$

Note that when fixed boundary conditions are assumed, $\varphi = \theta|_{x=L}$, and thus the rotation equation is not needed. But when pinned conditions are assumed, the equation of motion of the deck has to be solved simultaneously with the one for the pipe.

Numerical Solution Formulation

The equation of motion for the lateral displacement is a nonlinear partial differential equation with time-dependent coefficients. It is solved numerically using 'ACSL' - Advanced Continuous Simulation Language. The method of solution is based on the finite difference approach embedded into 'ACSL' as shown in a paper by Mitchell and Gauthier [11].

The length of the pipe is divided into N parts, as shown in Fig. 1.8, resulting in N nonlinear coupled ordinary differential equations which are then integrated in time using 'ACSL'. In this formulation, the boundary conditions are applied naturally at the ends of the space dimension.

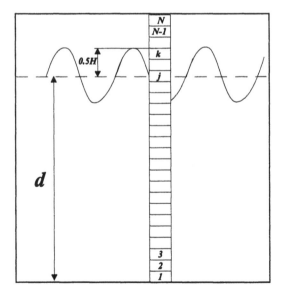

1.8 Discretization of the Submerged Pipe.

The partial derivatives with respect to x, for the central parts of the pipe, are approximated using central differences, the upper end via backward differences and the lower end using forward differences. The central difference approximation is of order $O\left[(\Delta x)^2\right]$, while the forward and backward approximations are of order $O\left[(\Delta x)\right]$, where Δx is

$$\Delta x = \frac{l}{N}. \tag{1.168}$$

The expressions for the central, forward and backward approximations can be found in Jaluria [12] pp. 254-256.

Central Difference Approximations

This approximation is used for $i = 3$ to $N - 2$:

$$
\begin{align}
y_x(i) &= \frac{y(i+1) - y(i-1)}{2\Delta x} \tag{1.169} \\
y_{xx}(i) &= \frac{y(i+1) - 2y(i) + y(i-1)}{(\Delta x)^2} \\
y_{xxx}(i) &= \frac{y(i+2) - 2y(i+1) + 2y(i-1) - y(i-2)}{2(\Delta x)^3} \\
y_{xxxx}(i) &= \frac{y(i+2) - 4y(i+1) + 6y(i) - 4y(i-1) + y(i-2)}{(\Delta x)^4} \\
y_{xt}(i) &= \frac{y_t(i+1) - y_t(i-1)}{2\Delta x} \\
y_{xtt}(i) &= \frac{y_{tt}(i+1) - y_{tt}(i-1)}{2\Delta x} \\
y_{xxtt}(i) &= \frac{y_{tt}(i+1) - 2y_{tt}(i) + y_{tt}(i-1)}{(\Delta x)^2}
\end{align}
$$

Backward Difference Approximations

This approximation is used for the last two nodes, $i = N - 2$ to N:

$$
\begin{align}
y_x(i) &= \frac{y(i) - y(i-1)}{\Delta x} \tag{1.170} \\
y_{xx}(i) &= \frac{y(i) - 2y(i-1) + y(i-2)}{(\Delta x)^2} \\
y_{xxx}(i) &= \frac{y(i) - 3y(i-1) + 3y(i-2) - y(i-3)}{(\Delta x)^3} \\
y_{xxxx}(i) &= \frac{y(i) - 4y(i-1) + 6y(i-2) - 4y(i-3) + y(i-4)}{(\Delta x)^4} \\
y_{xt}(i) &= \frac{y_t(i) - y_t(i-1)}{\Delta x} \\
y_{xtt}(i) &= \frac{y_{tt}(i) - y_{tt}(i-1)}{\Delta x}
\end{align}
$$

$$y_{xxtt}(i) = \frac{y_{tt}(i) - 2y_{tt}(i-1) + y_{tt}(i-2)}{(\Delta x)^2}$$

Forward Difference Approximations

This approximation is used for the first two nodes, $i = 1$ to 2 :

$$y_x(i) = \frac{y(i+1) - y(i)}{\Delta x} \tag{1.171}$$

$$y_{xx}(i) = \frac{y(i+2) - 2y(i+1) + y(i)}{(\Delta x)^2}$$

$$y_{xxx}(i) = \frac{y(i+3) - 3y(i+2) + 3y(i+1) - y(i)}{(\Delta x)^3}$$

$$y_{xxxx}(i) = \frac{y(i+4) - 4y(i+3) + 6y(i+2) - 4y(i+1) + y(i)}{(\Delta x)^4}$$

$$y_{xt}(i) = \frac{y_t(i+1) - y_t(i)}{\Delta x}$$

$$y_{xtt}(i) = \frac{y_{tt}(i+1) - y_{tt}(i)}{\Delta x}$$

$$y_{xxtt}(i) = \frac{y_{tt}(i+2) - 2y_{tt}(i+1) + y_{tt}(i)}{(\Delta x)^2}$$

A partially submerged pipe is subjected to different forces along its length, that is, wave forces for the submerged part, wave slamming forces for the intermediate part and wind forces for the exposed part of the pipe. Therefore, different equations have to be solved for each part. To do so, the parts of the pipe which are in the region between the mean water level d and $d + \frac{1}{2}H$ are found. Since the pipe is discretized, the following relations hold,

$$d = \sum_{j=1}^{j} \frac{\Delta x}{\sqrt{1 + y_x^2(i)}} = \sum_{j=1}^{j} \frac{l}{N\sqrt{1 + y_x^2(i)}}, \tag{1.172}$$

and

$$d + \frac{1}{2}H = \sum_{j=1}^{k} \frac{\Delta x}{\sqrt{1 + y_x^2(i)}} = \sum_{j=1}^{k} \frac{l}{N\sqrt{1 + y_x^2(i)}}, \tag{1.173}$$

where j is the last part which is submerged below mean water level d, and k is the last part submerged below $d + \frac{1}{2}H$.

The general equation for the ith part of the pipe used in the 'ACSL' code is

$$y_{tt}(i) = \frac{-F_{dyn}(i) + sw_1(Q_D(i) + Q_M(i))}{M_{eq}(i)} \tag{1.174}$$

$$+ \frac{(1 - sw_1)Q_w(i) + sw_2 Q_s(i)}{M_{eq}(i)},$$

where $F_{dyn}(i)$ is the dynamic force given by the left hand side of equation (1.166) with $y_{tt} = 0$. $Q_D(i)$, $Q_M(i)$, $Q_w(i)$, and $Q_s(i)$ are the drag, inertia, wind and wave slamming generalized forces respectively, given in equations (1.148), (1.150), (1.153) and (1.156). sw_1, sw_2 are switches that are set to $0, 1$ depending on the value of i. The equivalent mass of the ith part of the pipe $M_{eq}(i)$ is

$$M_{eq}(i) = \frac{1}{4}\pi D^2(i)\left(\rho_T + C_A\rho\right).$$ (1.175)

Once j and k are found, the following is applied; For $n = 1$ to j, the full dynamic equation with external forces only due to wave applies. For $n = j + 1$ to k, again the full dynamic equation with external forces due to waves and slamming waves applies. Finally, for $n = k + 1$ to N, the added mass and buoyancy terms are set to zero and external forces are only due to wind.

Summary

The nonlinear equations of motion for a compliant offshore structure such as an articulated tower, a *tension leg platform* and a *riser*, are derived. It is assumed that the deck motion and the cable/column attached to it are coupled, thus two equations are derived. Nonlinearities due to geometry and external forces are included in the analysis. The equations are general enough so that the dynamic motion of all three compliant structures mentioned above can be predicted for several boundary conditions. The governing equations can include, among other things,

- Earthquake excitation
- Non-constant internal fluid flow
- Fixed or pinned boundary conditions
- Random and/or deterministic wave and wind loads.

The corresponding equations for each type of compliant structure can be easily formulated by specializing the parameters in the equations of motion, (1.166) and (1.167) and then solved by the formulated numerical method . This is accomplished in the subsequent chapters.

2
Articulated Towers

The articulated tower consists of a vertical column to which a buoyancy chamber is attached near the water surface and to which a ballast is usually added near the bottom. The tower is connected to the sea-floor through a universal joint connected to a base. The tower itself may be either a tubular column or a trussed steel latticework. The structure's fundamental frequency is designed to be well below those wave frequencies with high amplitudes. Articulated towers are typically designed for water depths of 100 to 500 m and are used as single point mooring or as loading terminals, control towers and early and/or full production facilities.

Design and Construction

Very few papers discussing the design and construction aspects of articulated towers are found. The first articulated tower ever built was designed in response to an industry call, in 1963, for innovative offshore structures. A full scale experimental structure, for a water depth of 330 ft, was constructed and installed in 1968. The tower remained on site for three years during which time many measurements were taken for a wide variety of weather conditions. This experiment demonstrated that the articulated tower concept can be utilized in the offshore industry.

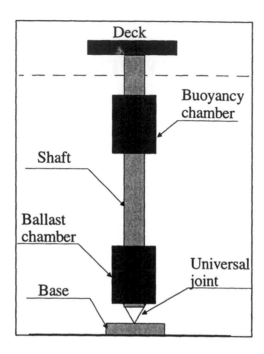

2.1 Schematic of an Articulated Tower.

Burns and D'Amorim (1977) [13] discussed the development, design and construction of two articulated towers that provide mooring facilities and house flow lines from subsea equipment to surface facilities. The towers were designed for a water depth of 420 *ft*. Environmental loads due to waves, current and wind were considered in the evaluation of the tower's dynamic response and the reaction forces in the base. The tower is constructed of the following major parts (see Fig. 2.1): *Base* - connects the tower to the sea-floor and keeps it from lifting or sliding. *Universal joint* - has two degrees of freedom, one can tilt 30^0 and the other 90^0, so that the tower can be constructed horizontally. It is designed to withstand horizontal loads of 1000 *kips*, and downward loads of 2000 *kips*. *Ballast chamber* - it is located above the joint and has a diameter of 32 *ft*. During towing it is pressurized to give buoyancy, but it is flooded when installed. *Shaft* - its diameter is 18 *ft*. It connects the ballast and the buoyancy chamber and it is full of water during service. *Buoyancy chamber* - the chamber has a 32 *in* diameter and provides the vertical force that keeps the tower upright.

Hays et al. (1979) [14] discussed the operation of an articulated oil loading tower in the North Sea at a water depth of 400 *ft*. The reasons for selecting the articulated tower concept were simplicity of design, that it could remain as an unmanned facility during loading operations, its superior underwater reliability and advantageous motion characteristics. The structure consists of a steel body which oscillates about a universal joint connected to the sea-floor via a concrete ballast. The head of the tower rotates, so when a tanker is moored to the tower, the orientation of the head is determined by the weather conditions. The tower was built, tested and, according to the authors, it fulfilled their expectations.

In two papers by Smith (1979) [15] and Smith and Taylor (1980) [16] , the applicability, function and performance of an articulated tower were examined. The construction program covered the following aspects; hydrodynamics, materials, economic assessments and interaction between the structure and the fluid. The tower was designed for a water depth of 250 *m*, and its predicted cost was approximately $700 *million*, while a conventional fixed structure would have cost on the order of $1.5 *billion*. An analysis of the response due to waves and wind was performed and the results were compared to a 1:64 scale model. A fairly good correlation between the analytical model and the experimental one was found. From the analysis and testing, the authors concluded the following;

- Articulated towers can perform a broad range of functions in offshore production.
- A good analytical model to predict the tower's response is important.
- A close collaboration with the oil industry, in order to address real problems, is needed.

In a paper by Butt et al. (1980) [17] , a large-scale test program for a concrete articulated tower was presented. The tests were planned to be performed in the vicinity of a research platform called 'Nordsee'. The aim of the test program was to demonstrate the technical feasibility of the CONAT (Concrete Articulated Tower) concept. The special features of this concept are the bottle-shaped concrete tower and the ball joint which creates the articulated connection. The problems of oscillating platforms and basic design steps were discussed. And finally the scheduled program tests were briefly discussed.

Naess (1980) [18] presented the results of an extensive scale 1:70 model test of an articulated tower. The model, 20 *m* in length, was built of steel and aluminium, to give the necessary strength, and polyurethane foam, to give the correct outer dimensions. The tests were done both with and without a tanker moored to it. The tests included the following measurements; wave elevation, pitch and roll of the tower were measured with accelerometers, axial and lateral (shear) forces were measured at the universal joint by strain gauges, bending moment near the buoyancy chamber and tension force at the mooring cable were also measured by strain gauges. The system was tested under a wind of 24 *m/s*, waves with significant wave height of 5.5 *m* and frequency 0.12 *Hz*, and current velocity 0.5 *m/s*. From the tests they found the natural period to be 58.6 *s* and the damping ratio $\zeta = 0.46$. The pitch angle was about 10^0. An attempt to measure the natural period in the presence of current failed because the free oscillations were immediately damped out, a phenomenon which is explained in Bar-Avi and Benaroya [19] .

An anonymous trade magazine paper (1990) [20] describes the world's largest single-point-mooring (SPM) terminal. It was constructed in December 1989 in the Timor Sea off Australia's northwest coast. The tower was designed to survive conditions of significant wave height up to 9*m*, wind velocity of 47 *m/s* and current velocity of 2 *m/s*. The operational environmental conditions were a significant wave height of 3*m*, wind velocity of 14.5 *m/s* and current velocity of 1 *m/s*. The main components of the SPM are the same as in any other articulated tower; a ballast, a universal joint, a tower and a mooring yoke.

Dynamic Response

Since articulated towers comply with the environmental forces, they can undergo large displacements. Therefore, the dynamic response of these kinds of structures is very important. Most of the studies considered the tower as a rigid body having a one or two angular degrees of freedom about a universal joint. Structures having multiple articulations in planar or 3D motion were also analyzed. Very few studies considered the tower as a flexible structure and those that did, used lumped mass or finite element methods and not the classical methods of continuum mechanics. The external forces considered by most studies were due to waves, current and wind. Linear wave theory was applied, and the forces were approximated by Morison's equation (1950) [2]. This subsection summarizes the literature on the dynamic response of articulated towers. It is divided to three sections:

- Single degree of freedom systems – The tower is assumed rigid and only planar motion was considered.
- Two degrees of freedom system – Rigid body and 3D analysis of motion and loads.
- Multiple articulation and flexible systems – Planar or 3D motion of multi–articulated towers or towers that were considered flexible.

Single Degree of Freedom Systems

Chakrabarti and Cotter (1978) [21] developed a mathematical model to analyze the dynamic response of a tower–tanker system. The tower was assumed rigid, connected to the tanker via a spring with stiffness K. Forces due waves, current and wind are considered collinear. First the static equilibrium state due to only current and wind was found. Then small perturbations about the equilibrium position were assumed in the formulation of the equation of motion. The tanker was assumed to have two degrees of freedom, one linear (surge), and the other angular (pitch). The equations for the tower and the ship were derived and coupled through the spring connecting the tower to the tanker,

$$I_t \gamma_{tt} + F_{Dt} + C_B \gamma + F_r l \cos(\theta + \varphi) = M_t e^{i(\varepsilon_1 - \sigma t)} \qquad (2.1)$$

$$m x_{tt} + F_{D1s} - F_r \cos\theta = F_t e^{i(\varepsilon_2 - \sigma t)} \qquad (2.2)$$

$$I_s \mu_{tt} + F_{D2s} + C_B \mu - F_r \left(H_s \cos\theta + \frac{L}{2} \sin\theta \right) = M_s e^{i(\varepsilon_3 - \sigma t)}, \qquad (2.3)$$

where γ, x and μ are the tower's deflection angle, and the ship's surge and pitch, respectively. F_r is the spring force which couples the tower to the ship, C_B is buoyancy term, F_{Dt}, F_{D1s}, and F_{D2s} are the drag forces proportional to the square of the relative velocity between the fluid and the structure, and σ is the wave frequency. The equations were solved numerically and the solution was compared to experimental results obtained from a model that was built with a scale of 1:48. Good correlation between the test results and theoretical predictions for small displacements was found. When soft spring–mass systems are considered, irregular waves produce a drift (static) force that the model did not predict.

In a later paper, Chakrabarti and Cotter (1979) [22] investigated the motion of an articulated tower fixed by a universal joint having a single degree of freedom. They assumed linear waves, small perturbations about an equilibrium position, and that the wind, current and wave are collinear. A linear equation was obtained by assuming a linear drag force and an analytical solution was obtained. The solution was then compared to experimental results, showing good agreement as long as the system is inertia dominant, and not drag dominant.

Kim and Lau (1981) [23] evaluated the response of an articulated loading platform in regular waves. The objective of the study was to develop a reliable technique to predict the loads and motion of the tower. The following assumptions have been made; rigid single degree of freedom body, linear drag force, small deflection angle, and deep water. In the derivation of the equation of motion, the tower was assumed to be in its upright position so that geometrical nonlinearities were not included. The linear equation of motion was then solved numerically and the solutions were compared to the experimental results presented by Naess [18] to show only qualitatively similarity, although using the same physical parameters. The authors concluded that a better model for the fluid forces, as well as not assuming deep water, will result in better predictions.

Muhuri and Gupta (1983) [24] investigated the stochastic stability of a buoyant platform. They used a linear single degree of freedom model,

$$x_{tt} + 2cx_t + (1 + G(t))x = 0, \qquad (2.4)$$

where x is the displacement, c is the damping coefficient and $G(t)$ is a stochastic time–dependent function due to buoyancy. It is assumed that $G(t)$ is a narrow–band random process with zero–mean. A criterion for the mean square stability is obtained from which the following results are found: for $c > 1$ the system is always stable, and for $0 < c < 1$ there are regions of stability and instability.

Thompson et al. (1984) [25] investigated the motions of an articulated mooring tower. They modeled the structure as a bilinear oscillator which consists of two linear oscillators having different stiffnesses for each half cycle,

$$mx_{tt} + cx_t + (k_1, k_2)x = F_0 \sin \omega t, \qquad (2.5)$$

where k_1 and k_2 are the stiffnesses for $x > 0$ and $x < 0$, respectively. The equation is solved numerically for different spring ratios and, as expected, harmonic and subharmonic resonances appeared in the response. A comparison between the response and experimental results of a reduced–scale model showed good agreement in the main phenomenon.

Chantrel and Marol (1987) [26] presented a study on a tanker moored to a single degree of freedom articulated tower. The objective of the study was to identify the relative importance of the different nonlinear terms in the equation of motion, especially the terms that cause subharmonic response. A few assumptions were made in deriving the equation of motion:

- Restoring moment due to buoyancy quadratic terms were neglected.
- The drag force due water velocity was neglected.
- Forces/moments were evaluated at the upright position of the tower.
- The force in the mooring cable was assumed to have a cubic form.

Applying these assumptions resulted in a nonlinear equation of motion that is linearized by assuming small perturbations about an equilibrium position, resulting in

$$u_{tt} + 2\gamma\omega_n u_t + \omega_n(1 + f_v \cos(\omega t + \varepsilon))u = \frac{M_0}{I} \cos(\omega t + \varepsilon). \qquad (2.6)$$

This equation is actually the Mathieu equation, and a stability analysis was performed to show an unstable region around the first natural frequency of the system. This region, as expected, gets smaller when linear damping is added to the system. Equation (2.6) was then solved numerically for regular and irregular waves having the Pierson–Moskowitz spectrum and the following conclusions were drawn:

- The subharmonic response is due to the nonlinear characteristic of the mooring cable's stiffness.
- The subharmonic response occurs for very specific environmental conditions.
- A region of parametric instability that depends on the system's damping was found.

Datta and Jain (1987) [27] , and (1990) [28] investigated the response of an articulated tower to random wave and wind forces. In the derivation of the single degree of freedom equation of motion the tower is discretized into n elements having appropriate masses, volumes and areas lumped at the nodes, and there is viscous damping. From their parametric study, the following was concluded:

- Nonlinearities such as large displacements and drag force do not influence the response when only wind force is considered.
- The random wind forces result in higher responses than do only wave forces.
- The root mean square response due only to wind forces varies in a linear fashion with the mean wind velocity.

In a later paper, Jain and Datta (1991) [29] used the same equation and the same method of solution to investigate the response due to random wave and current loading. The wave loadings (drag, inertia and buoyancy) are evaluated via numerical integration. The following results were obtained from the parametric study:

- The dynamic response is very small since its fundamental frequency is well below the wave's fundamental frequency.
- Nonlinear effects (drag force, variable buoyancy) have considerable influence on the response.
- Current velocity has a large influence on the response.

Virgin and Bishop (1990) [30] studied the domains of attraction (catchment regions) for a single degree of freedom articulated tower connected to a tanker. This was done using numerical techniques based on Poincar' mapping ideas. A basic bi–linear oscillator model was assumed, the equation of motion was the same as equation (2.5). This equation can exhibit multiple solutions, but in the example solved, the coefficients (stiffness and mass) were chosen so that only two solutions may coexist, depending on initial conditions; harmonic and 4th order subharmonic. The equation was solved numerically and it was shown that a domain of attraction could be found.

Choi and Lou (1991) [31] have investigated the behavior of an articulated offshore platform. They modeled it as an upright pendulum having one degree of freedom, with linear springs at the top having different stiffnesses for positive and negative displacements (bilinear oscillator). The equation of motion is simplified by expanding nonlinear terms into a power series and retaining only the first two terms. They assumed that the combined drag and inertia moment is a harmonic function. The discontinuity in the stiffness is assumed to be small, and thus replaced by an equivalent continuous function using a least–squares method to get the following Duffing equation

$$I\theta_{tt} + c\theta_t + k_1\theta + k_2\theta^2 + k_3\theta^3 = M_0 \cos \omega t, \qquad (2.7)$$

where k_1, k_2, and k_3 are spring constants depending on buoyancy, gravity and the mooring lines. The equation of motion is solved analytically and numerically, and stability analysis is performed. The analytical solution agrees very well with the numerical solution. The main results of their analyses are that as damping decreases, jump phenomena and higher subharmonics occur, and chaotic motion occurs only for large waves and near the first subharmonic (excitation frequency equals twice the fundamental frequency); the system is very sensitive to initial conditions.

Gottlieb et al. (1992) [32] analyzed the nonlinear response of a single degree of freedom articulated tower. In the derivation of the equation, the expressions for the buoyancy moment arm, added mass term, and drag and inertia moments are evaluated along the stationary upright tower position and not at the instantaneous position of the tower. The solution shows a jump phenomenon in the primary and secondary resonances.

Gerber and Engelbrecht (1993) investigated the response of an articulated mooring tower to irregular seas. It is an extension of earlier work done by Thompson et al. (1984) [25] where the tower is modeled as a bilinear oscillator. The random forcing function $F(t)$ is assumed to be the sum of a large number of harmonic components, each at different frequencies, a procedure similar to that proposed by Borgman (1969) [3] . The equation is then solved analytically since it is linear for each half cycle. The solution is obtained for different cases; linear oscillator (both stiffnesses are the same), bilinear oscillator, and for the case of impact oscillator (a rigid cable) in which oscillation can occur only in one half of the cycle. For future study they suggest inclusion of nonlinear stiffness and/or using a different spectrum to describe the wave height.

Bar–Avi and Benaroya (1996) [33] investigated the nonlinear response of a single degree of freedom articulated tower. The equation of motion was derived via Lagrange's equation. Nonlinearities due to geometry and wave drag force are considered. A combined wave and current field, Coulomb friction force, and vortex shedding force are included in the analysis. The governing equation of motion is

$$J(\theta)\theta_{tt} + C\theta_t + M_{gb}(\theta, t) = M_{fl}(\theta, t) - M_{fr}(\theta), \qquad (2.8)$$

where $J(\theta)$ is a position–dependent moment of inertia that includes added mass terms, C is the structural damping coefficient, $M_{gb}(\theta, t)$ is a time and position–dependent moment due to gravity and buoyancy, $M_{fl}(\theta, t)$ is the fluid moment due to inertia, drag and vortex shedding force, and $M_{fr}(\theta)$ is the friction moment. The influence on the response of current velocity and direction, significant wave height and frequency, and damping mechanism was analyzed. The following observations were made:

- The equilibrium position is proportional to the product of the current velocity squared and the drag coefficient.
- The highest response is when the current direction is perpendicular to the wave.
- The response to sub/superharmonics and harmonic excitation demonstrate beating.

- For most excitation frequencies, the response is quasiperiodic, but for certain frequencies chaotic behavior was observed.
- Damping (friction, structural) has a stabilizing effect.

A simplified equation for a single degree of freedom articulated tower was presented in Bar–Avi [34] . The equation was derived using the Taylor expansion of the fully nonlinear equation derived in [33] . Terms of second power or less were retained and the solutions of both the fully nonlinear and the simplified equations were compared. From the comparison it was found that the simplified equation predicts the tower's response very well over a broad regime of behavior. Analytical expressions for the natural frequency and equilibrium position due to current were presented. It was also shown that ocean current causes an additional damping mechanism in the system that can be approximated by $\frac{1}{3}C_D\rho D d^3 U_c \theta_t$. This result agrees with the experimental results presented by Naess (1980) [18] .

Two Degree of Freedom Systems

Kirk and Jain (1977) [35] investigated the dynamic response of a two degree of freedom articulated tower to non–colinear waves and current. The two equations of motion were obtained via Lagrange's equation. The waves were assumed linear with the current modifying the frequency and amplitude. Forces due to buoyancy, wave drag and inertia, and added mass were considered. The equations were solved numerically, and the influence of drag coefficient and wave direction was analyzed. From the solution they concluded that:

- Higher drag coefficients result in lower response.
- The maximum deflection occurs when the current and the waves are in the same direction.

When vortex shedding forces are included the last conclusion is not correct as shown in Bar–Avi and Benaroya (1996) [33] .

Olsen et al. (1978) [36] evaluated the motion and loads acting on a single–point mooring system. The tower was modeled as a rigid body connected to the sea–floor via a universal joint. The equation for the tower and the tanker were derived separately. To derive the equations of motion for the tower, it was divided into N elements having two degrees of freedom each; a horizontal and vertical displacement and the forces due to wave, current and wind were evaluated at each element. Hence, $2N$ nonlinear differential equations were found. The solution of these give the low frequency motion of the tanker as well as the tower's response. The high frequency motion of the tanker was assumed to be unaffected by the fact that the tanker is moored. Therefore the tanker's high–frequency amplitudes were calculated independently from the low frequency, and then the responses were added together. The equations were solved numerically and compared to test results to show a reasonably good correlation, but according to the authors, a more accurate model should be developed. The effect of the tanker on the surrounding wave field was also investigated to find a 0–10% change in the wave velocity and acceleration. The effect of these changes on the response was not investigated.

Chakrabarti and Cotter (1980) [37] investigated transverse motion, the motion perpendicular to the horizontal velocity. The tower pivot is assumed to have two angular degrees of freedom and is taken to be frictionless. It is also assumed that the motion is not coupled, so the in–line solution is found (the same as in the previous paper), from which the relative velocity between the tower and the wave is obtained. The lift force (in the transverse direction) can then be determined and the linear equation of motion is solved analytically and compared to experimental results. The comparison shows good agreement, especially when the drag terms are small.

Schellin and Koch (1985) [38] calculated the dynamic response due to waves and compared results with model tests. The calculation of the response was done for three different sets of fluid coefficients; coefficients that depend on the wave period, coefficients selected from experimental data and coefficients that are calculated using diffraction theory. The tower was assumed rigid and connected to the sea–floor via a two degree of freedom universal joint. Forces due to wind, wave and current as well as nonlinearities due to geometry and wave drag force are considered. The tower was divided into N elements for which the following force was found;

$$\mathbf{F}_i = C_{Mi}\dot{\mathbf{u}}_{wi} + C_{Di}\mathbf{u}_{wi}\left|\mathbf{u}_{rel}\right| + \mathbf{F}_{Bi} + \mathbf{F}_{Gi} + \mathbf{F}_{Wi} - \mathbf{m}_i\frac{d^2\mathbf{r}_i}{dt^2} - C_{Di}\frac{d\mathbf{r}_i}{dt}\left|\mathbf{u}_{rel}\right|, \quad (2.9)$$

where $\mathbf{F}_i$ is the vector of the total force acting on the element due to fluid inertia force $C_{Mi}\dot{\mathbf{u}}_{wi}$, fluid drag force $C_{Di}\mathbf{u}_{wi}\left|\mathbf{u}_{rel}\right|$, buoyancy, gravity and wind forces $\mathbf{F}_{Bi}, \mathbf{F}_{Gi}, \mathbf{F}_{Wi}$, inertia force due to tower's acceleration $\mathbf{m}_i\frac{d^2\mathbf{r}_i}{dt^2}$, and drag force due to tower velocity $C_{Di}\frac{d\mathbf{r}_i}{dt}\left|\mathbf{u}_{rel}\right|$. Summing all forces on each element and multiplying by the moment arm leads to the equations of motion that were solved numerically for an idealized tower consisting of a series of circular cylinders. The numerical solutions were compared to experimental results of a model which has been built to a scale of 1:32.75 and the following conclusions have been drawn:

- Proper choices of the drag and the added mass coefficients results in good correlation between the theoretical and test results for the tower's deflection and horizontal reaction force on the universal joint.
- The correlation of the vertical reaction force is not good.
- The added mass coefficient has a predominant effect on the dynamic response and the drag coefficient almost none.
- The theoretical results for waves having small period (high frequency) do not correlate as well as results for waves with large period.

Liaw (1988) [39] studied the nonlinear dynamic response of articulated towers subjected to regular waves. The tower is defined by a second order differential equation of a simple oscillator,

$$x_{tt} + \omega^2 x = \alpha \dot{U} + \alpha\beta\left|U - \dot{x}\right|(U - \dot{x}), \quad (2.10)$$

where α and β are constants depending on the fluid coefficients and tower dimensions. The nonlinearities are due to large displacement and velocity of the tower that are coupled to the wave force. The equation was solved numerically and the following conclusions were drawn:

- In addition to superharmonic resonances of order $\frac{1}{2}$, there were also superharmon-

40   Chapter 2

ics of orders $\frac{1}{3}, \frac{1}{5}, \frac{1}{7}...$ and so on.

- Chaotic motion was found in certain frequency regions.
- Before and within the chaotic regions, bifurcation behavior was identified.

In a later paper, Liaw et al. (1989) [40] formulated the equations of motion for a two degree of freedom articulated tower using Lagrange's equations, and then solved and analyzed the large motion of the structure. The equation was solved for three cases. First, the static equilibrium inclination of the tower due to current was obtained. Next, the response due to linear waves with height of 3 m and period of 17 s was evaluated. Finally, the previous waves along with orthogonal current were applied and the solution was found. All three cases were compared to the solution obtained by Leonard and Young (1985) [41], who used a finite element method, and the results matched quite well.

In a paper from 1992, Liaw et al. [42] showed that the superharmonic phenomenon, which occurs in articulated towers, is due to the coupling between the wave force and the structure. They used the equations that were developed in their previous paper [40], but reduced them from 2 degrees of freedom to a single degree of freedom system. The equation was solved numerically and harmonic and subharmonic responses were obtained. The following observations were made:

- The amplitude of the response in the superharmonic region can be as high as the one in the harmonic region.
- The initial conditions determine the final steady state response.

Similar results for a single degree of freedom model were obtained in a study presented by Bar–Avi and Benaroya (1996) [33], although, for a two degree of freedom system (see [43]) it was found that the superharmonic response is not as pronounced as in the single degree of freedom model.

Bar–Avi and Benaroya (1995) [19], [43], investigated the response of a two degree of freedom articulated tower to deterministic loading. The nonlinear differential equations of motion were derived using Lagrange's equations. The tower was assumed to have the same dynamic properties as an upright spherical pendulum with additional effects and forces:

- Coulomb friction in the pivot (hinge)
- Structural viscous damping
- Drag fluid force due to waves and current, and wind coupled to the structure
- Inertia, buoyancy and added mass fluid forces
- Vortex shedding loads due to waves and wind
- Wave slamming that was modeled as a periodic impulsive force
- Gyroscopic moments due to the rotation of the earth (Coriolis acceleration).

All fluid forces due to waves, current, and wind are determined at the instantaneous position of the tower, resulting in two, highly nonlinear, coupled, ordinary differential equations with time–dependent coefficients, with rotation angle φ, and deflection angle θ,

$$J_{eff}\theta(\theta)\theta_{tt} + C\theta_t + I_g(\theta)\varphi_t^2 + M_{gb}\theta(\theta) = M_T$$

$$J_{eff}\varphi(\theta)\varphi_{tt} + C\varphi_t + I_g(\theta)\varphi_t\theta_t = M_T$$
$$M_T = M_{fl}\varphi(\theta, t) + M_{sm}\varphi(\theta, t) + M_w\varphi(\theta) - M_{fr}\varphi(\theta, t). \qquad (2.11)$$

The equations of motion were numerically solved and the following observations were made:

- An analytical expression for the equilibrium position due to current and wind was found.
- The response due to wave slamming is very small since an impulsive force is attenuated when the pulse duration is shorter than the system's fundamental period; this is the case here.
- Wind loads and current loads affect the equilibrium position of the tower.
- The Coriolis acceleration force has a small but important influence on the response since it causes a coupling so that planar motion is not possible under real conditions.
- The regions in which the beating phenomenon occurs are very small and not as pronounced as in a single degree of freedom system.
- Due to the system's nonlinear behavior, chaotic regions exist.

Later, Bar–Avi and Benaroya (1995) [44], [45], analyzed the response of a two degree of freedom tower, where key parameters were taken to be random variables. The wave height, drag, inertia and lift coefficients, and Coulomb friction coefficient were assumed to be random uniformly distributed variables. The nonlinear differential equations of motion were numerically solved and Monte–Carlo simulations were performed to evaluate the average response and the standard deviation. It was found that the standard deviation for the rotation angle is larger than that of the deflection angle. The value of the friction coefficient has a very small influence on the average response, unlike the wave height and the drag coefficient.

Multiple Articulations and Flexible Systems
In a paper by Jain and Kirk (1981) [46], a double articulated offshore structure subjected to waves and current loading was analyzed. They assumed four degrees of freedom, two angular degrees for each link. The equations of motion were derived using Lagrange's equations. In deriving the equations of motion the following assumptions were made: drag and inertia forces tangent to the tower are negligible, and the wave and current velocities are evaluated at the upright position (small angles assumption). The linearized equations were solved to find the natural frequencies of the system and then numerically solved to find the response due to colinear and non–colinear current and wave velocities. They found that when the wave and the current velocities are colinear, the response of the top is sinusoidal, while for non–colinear velocities the response is a complex three dimensional whirling oscillation.

Seller and Niedzwecki (1992) [47] investigated the response of a multi–articulated tower in planar motion (upright multi–pendulum) to account for the tower flexibility. The restoring moments (buoyancy and gravity) were taken as linear rotational springs between each link, although the authors state that nonlinear springs are more adequate for this model. Each link is assumed to have a different cross section and density. The equations of motion are derived using Lagrange's equations, in which the generalized coordinates are the angular deflections of each link. The equations in matrix form are

$$[M]\{\theta_{tt}\} + [K]\{\theta\} = \{Q\}, \tag{2.12}$$

where $[M]$ is a mass matrix that includes the actual mass of the link and added mass terms, while the stiffness matrix $[K]$ includes buoyancy and gravity effects. Damping and drag forces are not included in the model. The homogeneous equations for a tri–articulated tower are numerically solved to study the effects of different parameters, such as link length, material density and spring stiffness, on the natural frequency of the system.

In two papers, one by Havery et al. (1982) [48] , and by McNamara and Lane (1984) [49] , a finite element method is used to calculate the response of a planar flexible multi–articulated tower. Examples of the response of single point mooring, bi–articulated and multi–articulated towers were presented. In order to derive the equation of motion, the displacement was decomposed into a rigid body motion and a deformed motion. Two coordinate systems were used. One fixed and the other attached to the tower's rigid body motion. The deformation was first expressed in the rotating system and then transformed into the fixed coordinate system in which the equation of motion was expressed for each element, to find

$$Mw_{tt} + Kw = Kw^{rb} + F, \tag{2.13}$$

where M is the mass matrix, K is the stiffness matrix, w, w^{rb} are the total and rigid body displacements, respectively, and F is the force due to wave and current, calculated by Morison's equation. For the random wave, the Pierson–Moskowitz spectrum was transformed into the time domain using Borgman's method [3] . The equations were solved numerically using a finite difference method in which artificial damping was introduced which, according to the authors, does not significantly influence the response. It was found that the finite element solution using 21 elements was stable up to a time step of 0.7 s. A solution for the same problem, based on numerical integration of the Lagrange's equations (not presented), was compared to the finite element solution and the results agreed exactly except for a few initial cycles. The method presented can be extended to more realistic problems such as two degree of freedom universal joints.

The objective of the paper by Leonard and Young (1985) [41] was to develop a solution method to evaluate the dynamic response of an articulated tower. The method is based on three dimensional finite elements. The tower was subjected to wave, current and nonlinearities due to geometry and drag force were included. The equations of motion are

$$[M]\{q_{tt}\} + [C]\{q_t\} + [K]\{q\} = \{F(t)\}, \tag{2.14}$$

where $[M]$, $[C]$, and $[K]$ are the mass, damping and stiffness matrices, $\{q_{tt}\}, \{q_t\}$, and $\{q\}$ are the generalized acceleration, velocity and displacement vectors and $\{F(t)\}$ is the generalized force vector due to wave and current. The response was evaluated numerically for steady current only and then for waves. The results were compared to those presented by Jain and Kirk (1981) [46] . From this comparison it was concluded that the three dimensional finite element method is adequate. For linear analysis it requires more time than other linear computer schemes, but when nonlinearities are included the method actually requires less time than others.

Sebastiani et al. (1984) [50] presented the design and dynamic analysis of a 1000 m single point mooring tower in the Mediterranean Sea. The tower consists of four slender columns, about 3 m in diameter each, connected via a universal joint. A buoyancy chamber is welded to the upper column, just underneath the deck. The tower is so flexible that several natural frequencies are within the range of the wave frequencies. Therefore, it operates in a resonance condition. The structure was modeled using finite elements, and forces due to wave, current and wind were considered. In the dynamic analysis that was performed for survival and station keeping conditions, irregular seas characterized by the Pierson–Moskowitz spectrum were considered. To gain a better understanding of the dynamic behavior, a model test program was launched. A model having a of scale of 1:107.5 was built and tested. The tests were performed for the structure alone in wind, current and waves, and then with a tanker moored to the structure. The test results showed reasonable agreement with the simulations for the maximum forces. As for the dynamic behavior, the theoretical predictions did not agree with the tests.

Hanna et al. (1988) [51] investigated the dynamic response of tri–articulated towers subjected to wave, wind and current. The tower geometry and dynamic characteristics were optimized such that tower periods fall outside the 5 to 20 s range, and reaction forces and weight are minimized. The model consisted of three rigid segments with different lengths and masses, and a total length of 3000 ft. Each segment had a single degree of freedom and they were connected via a rotational spring. Thus, three linear ordinary differential equations were obtained for small angles

$$[M]\{x_{tt}\} + [K]\{x\} = \{F\}, \qquad (2.15)$$

where $[M]$,and $[K]$ are 3×3 mass and stiffness matrices, respectively, $\{F\}$ is the forcing vector due to wave, wind and current, and $\{x\}$ is the displacement vector. Equation 2.15 was used to determine the static stability due to offsets of the deck weight. Values for segment length, weight and joint stiffness were found for the highest critical load. To analyze the dynamic response and the stresses, large angular deflections were considered. The tower was divided into N elements each having a single degree of freedom. Nonlinearities due to geometry and drag forces were included, resulting in

$$[M]\{u_{tt}\} + [C]\{u_t\} + [K]\{u\} = \{P(t, u, u_t)\}, \qquad (2.16)$$

where $\{P(t, u, u_t)\}$ is the vector of the forces due to waves and colinear current approximated by Morison's equation, and due to static wind loads. Numerical solutions were obtained for deterministic and irregular waves having the Pierson–Moskowitz spectra. From the analysis it was concluded that compliant towers with multiple articulations provide an attractive concept to optimize the dynamic response without penalizing the structure's weight. Furthermore, the method of analysis can be utilized for 3D structures and also other similar compliant towers with multiple articulations.

Helvacioglu and Incecik (1988) [52] described analytical models to predict the dynamic response of a single and bi–articulated tower subjected to waves and wind. The analytical solutions were compared to test measurements. The effects of changes in the buoyancy position, joint location and deck weight on the bi–articulated tower response were studied. In both models, planar motion was assumed, and although it wasn't mentioned in the paper, fluid drag forces were not included, and therefore simple equations were derived that resulted in simple analytical solutions. In both models the equations were simple oscillators with damping subjected to harmonic forces,

$$[M]\{\theta_{tt}\} + [C]\{\theta_t\} + [K]\{\theta\} = \{M_F\}\sin\omega t, \qquad (2.17)$$

where $[M]$, $[C]$, and $[K]$ are scalars for a single articulation or 2×2 matrices for the bi–articulated system. From the parametric study it was found that the buoyancy tank position has a significant effect on the natural frequencies. According to the authors, the mathematical model for the bi–articulated tower correlated reasonably with the test results.

Active control of offshore articulated towers was discussed in a paper by Yoshida et al. (1988) [53] . A preliminary attempt was to control the dynamic response of an articulated tower subjected to regular waves. Two models were used; one was a rigid body having a single degree of freedom, and the other was a flexible tower fixed at the bottom. The control scheme was expressed as a combination of feedforward control based on the disturbance and a feedback control. The feedback control copes with the higher order noise remaining after the compensation of the feedforward control. Two feedforward control schemes were discussed. One is to compensate for the whole wave force acting on the structure, while the other was on–off control to compensate for the principle Fourier components of the wave force. The simulation results for both models showed that the response of the controlled structure was reduced to about 30% of those of the uncontrolled system.

In a later paper, Yoshida and Suzuki (1989) [54] discussed the experimental results of the response of an actively controlled tri–articulated tower. The application of active control to offshore structures is advantageous, increasing strength (stiffness) and reducing weight. The structure can be artificially stiffened and damped by means of active control according to the environmental conditions. Ultrasonic sensing systems were used to measure the deflection of each segment of the tower. The data from the measuring system was processed and the signals for the controllers were obtained. The control force was generated by thrusters which were built into each segment. Optimal control was applied to several cases:

- A neutral model, in which the buoyancy and gravity forces are equal, was controlled. The response of the model against an imposed displacement was controlled. The thrusters had a phase delay, and therefore vibrations in high frequency could not be controlled.

- An unstable model, in which the buoyancy force was less than the gravity force was controlled. In this case the structure was stabilized but again high frequency vibrations could not be controlled.

- Static deflection due to current was controlled successfully, but with large deflection angle the high gain necessary to control the structure caused some instability.

Ganapathy et al. 1990 [55] developed a general finite element program for the analysis of the nonlinear statics and dynamics of articulated towers. The tower was modeled as a three–dimensional beam element, which includes axial shear and bending deformations. The equations of motion have the standard finite element formulation and Linear wave theory was assumed and the wave force was evaluated via Morison's equation. The equations were numerically solved and the effects of the water depth, buoyancy force magnitude and position, and wave and current loads were investigated and the following conclusions were drawn:

- For moderate water depth (100 m), the maximum bending moment occurs at the position of the chamber, while for deep water (300 m), the maximum bending moment can occur at the mid span or at the buoyancy chamber, depending on the chamber's position.
- Larger buoyancy forces cause a decrease in the tower's deflection and an increase in the bending moment.
- Current load has a significant effect on the deflection and the bending moment.
- There is a nonlinear relation between the total force and the deflection.

Mathisen and Bergan (1991) [56] outline a general approach to large displacement static and dynamic analysis of an interconnected rigid and deformable multibody system submerged or floating in water. The system's equations of motion were generated by combining the equation of motion derived for each subsystem, which can be either rigid or deformable. The investigation is based on the Lagrangian description of motion in which the current coordinate of a material point is described in terms of its initial material point and time. The equation of motion of each part was derived using variational methods, and then combined with a nonlinear finite element displacement formulation. The formulation was applied to a bi–articulated tower, and the purpose was to find the response of the top of the platform, as well as to evaluate the distribution of the axial force and bending moment along the tower. The equations were solved for a deterministic wave height of 30 m with a period of 30 s, and irregular (random) waves having the Jonswap spectrum.

Rigid Body Versus Elastic Body Dynamics

Before moving towards a full analysis of a rigid articulated tower, the assumption of rigid body motion will be shown to be valid for this kind of structure in a practical application. The natural frequency and the modes of vibration are the key factors in this proof. It will be shown that the first mode of the system is that of a rigid pendulum, and has a much lower frequency and higher amplitude than the second mode, which corresponds to the first elastic mode. The first elastic frequency is also very high compared to the range of wave excitation frequencies where most of the energy resides, and hence is not influenced by the wave forces. In order to find an analytical solution for the natural frequencies the following assumptions are made:

- The buoyancy force acts at the top of the tower. This is a reasonable assumption for the articulated tower large scale dynamics.
- The mass of the deck is much larger than the total mass of the tower, $\rho_T \pi D^2 l / 4 \ll M$.
- The length of the tower is much larger than its diameter, $D \ll l$.
- Small displacements, $y/l \ll 1$, occur about equilibrium position $y = 0$.

- External forces are set to zero, $Q = 0$.

Applying the above assumptions in equation (1.166) yields the following partial differential equation of motion,

$$y_{xxxx} + N y_{xx} + \tilde{m} y_{tt} = 0, \tag{2.18}$$

with boundary conditions

$$\text{at } x = 0, \quad \left\{ \begin{array}{l} y = 0 \\ y_{xx} = 0, \end{array} \right. \tag{2.19}$$

$$\text{at } x = l, \quad \left\{ \begin{array}{l} y_{xx} = 0 \\ y_{xxx} + N y_x = \tilde{M} y_{tt}, \end{array} \right. \tag{2.20}$$

where

$$N = \frac{64g}{E\pi D^4} \left(M - \frac{1}{4}\rho\pi D^2 d \right) \tag{2.21}$$

$$\tilde{m} = \frac{16g}{E D^2} \left(\rho_T + C_A \rho \right) \tag{2.22}$$

$$\tilde{M} = \frac{64M}{E\pi D^4}. \tag{2.23}$$

To solve equation (2.18), separation of variables with harmonic solution in time is assumed,

$$y = \bar{y}(x) e^{-i\omega t}. \tag{2.24}$$

Substituting equation (2.24) into equation (2.18) and rearranging terms leads to the equation for the displacement

$$\bar{y}_{xxxx} + N \bar{y}_{xx} - \tilde{m}\omega^2 \bar{y} = 0, \tag{2.25}$$

with boundary conditions

$$\text{at } x = 0, \quad \left\{ \begin{array}{l} \bar{y} = 0 \cdot \\ \bar{y}_{xx} = 0, \end{array} \right. \tag{2.26}$$

$$\text{at } x = l, \quad \left\{ \begin{array}{l} \bar{y}_{xx} = 0 \\ \bar{y}_{xxx} + N \bar{y}_x + \tilde{M}\omega^2 \bar{y} = 0. \end{array} \right. \tag{2.27}$$

The solution of equation (2.25) is of the form

$$\bar{y} = \bar{A} e^{\lambda x}. \tag{2.28}$$

Substituting equation (2.28) into equation (2.25) and rearranging leads to the characteristic equation

$$\lambda^4 + N\lambda^2 - \tilde{m}\omega^2 = 0. \tag{2.29}$$

Thus, the solution for λ is

$$\lambda_{1,2} = \pm\sqrt{-\frac{1}{2}(N - \beta)} = \pm\sqrt{\frac{1}{2}(\beta - N)} = \pm\alpha_1 \tag{2.30}$$

$$\lambda_{3,4} = \pm\sqrt{-\frac{1}{2}(N+\beta)} = \pm i\sqrt{\frac{1}{2}(\beta+N)} = \pm i\alpha_2, \qquad (2.31)$$

where β is

$$\beta = \sqrt{N^2 + 4\tilde{m}\omega^2}. \qquad (2.32)$$

Substituting $\lambda_1, \lambda_2, \lambda_3$, and λ_4 into equation (2.28) yields

$$\bar{y} = \bar{A}e^{\alpha_1 x} + \bar{B}e^{-\alpha_1 x} + \bar{C}e^{i\alpha_2 x} + \bar{D}e^{-i\alpha_2 x}. \qquad (2.33)$$

Rearranging this solution leads to the general solution of equation (2.25),

$$\bar{y} = A\sinh\alpha_1 x + B\cosh\alpha_1 x + C\sin\alpha_2 x + D\cos\alpha_2 x. \qquad (2.34)$$

Applying the first two boundary conditions, B and D are related by

$$\bar{y}(0) = 0 \Longrightarrow B + D = 0 \qquad (2.35)$$
$$\bar{y}_{xx}(0) = 0 \Longrightarrow \alpha_1^2 B - \alpha_2^2 D = 0, \qquad (2.36)$$

which leads to

$$B = D = 0. \qquad (2.37)$$

The boundary conditions at $x = l$ yields two homogeneous equations for the remaining constants,

$$\begin{pmatrix} Z_{11} & Z_{12} \\ Z_{21} & Z_{22} \end{pmatrix} \begin{Bmatrix} A \\ C \end{Bmatrix} = \begin{Bmatrix} 0 \\ 0 \end{Bmatrix}, \qquad (2.38)$$

where

$$Z_{11} = \alpha_1^2 \sinh\alpha_1 l \qquad (2.39)$$
$$Z_{12} = -\alpha_2^2 \sin\alpha_2 l \qquad (2.40)$$
$$Z_{21} = \alpha_1^3 \cosh\alpha_1 l + N\alpha_1 \cosh\alpha_1 l + \tilde{M}\omega^2 \sinh\alpha_1 l \qquad (2.41)$$
$$Z_{22} = -\alpha_2^3 \cos\alpha_2 l + N\alpha_2 \cos\alpha_2 l + \tilde{M}\omega^2 \sin\alpha_2 l. \qquad (2.42)$$

Setting the determinant of the matrix in equation (2.38) to zero yields the characteristic equation from which the natural frequencies can be found,

$$\omega \left[\tilde{m} \left(\alpha_2^3 \coth(\alpha_1 l) - \alpha_1^3 \tan(\alpha_2 l) \right) + \tilde{M}\omega\beta \right] = 0. \qquad (2.43)$$

The solution of the characteristic equation (2.43) yields the set of natural frequencies ω_n of the system with $n = 1, 2.., \infty$. By observation it can be seen that the first natural frequency of the system equals zero, $\omega_1 = 0$, corresponding to rigid body motion. The mode associated with this natural frequency is found by substituting $\omega = 0$ into equation (2.25), which results in the partial differential equation for the rigid body motion,

$$\bar{y}_{xxxx} + N\bar{y}_{xx} = 0, \qquad (2.44)$$

with the following boundary conditions,

$$\text{at } x = 0, \quad \begin{cases} \bar{y} = 0 \\ \bar{y}_{xx} = 0, \end{cases} \qquad (2.45)$$

$$\text{at } x \;=\; l, \quad \begin{cases} \bar{y}_{xx} = 0 \\ \bar{y}_{xxx} + N\bar{y}_x + \tilde{M}\Omega^2 \bar{y} = 0, \end{cases} \tag{2.46}$$

where Ω is the rigid body natural frequency.

The general solution of equation (2.44) is

$$\bar{y} = A_1 + A_2 x + A_3 \sin\sqrt{N}x + A_4 \cos\sqrt{N}x, \tag{2.47}$$

which can be obtained by solving $z_{xx} + Nz = 0$ for z, where $z = \bar{y}_{xx}$, and then integrating z twice to obtain $\bar{y}$. This solution consists of a rigid body motion and an elastic motion. Applying the first three boundary conditions to equation (2.47) leads to

$$A_1 = 0, A_3 = 0, A_4 = 0. \tag{2.48}$$

Now, only the rigid body motion is left. The last boundary condition leads to the following equation,

$$A_2 N + A_2 \tilde{M}\Omega^2 l = 0, \tag{2.49}$$

from which the nontrivial solution leads to the natural frequency of the tower as a rigid body Ω,

$$\Omega^2 = \frac{-N}{\tilde{M}l}, \tag{2.50}$$

where N is a negative number. Substituting the definition for N and $\tilde{M}$ from equations (2.21) and (2.23) yields

$$\Omega^2 = \frac{g\left(\frac{1}{4}\rho\pi D^2 d - M\right)}{Ml}. \tag{2.51}$$

From equation (2.51) it can be seen that the condition for stability is

$$\frac{1}{4}\rho\pi D^2 d - M > 0, \tag{2.52}$$

which means that for stability, the buoyancy force has to be larger than the gravity force.

Equation (2.50) is the natural frequency of a pendulum. For a simple pendulum, $N = -\tilde{M}g$, and thus the following well–known expression is found

$$\Omega^2_{sp} = \frac{g}{l}. \tag{2.53}$$

Hence the solution to equation (2.44) becomes

$$y = A_2 x \sin\Omega t, \tag{2.54}$$

which describes the rotational motion of a rigid pendulum about the hinge. A numerical solution of equation (2.18) is compared to the semi-analytical solution from which the following is obtained.

The first two natural frequencies of the system are shown in Fig. 2.2. The lower natural frequency is the rigid body frequency $\Omega = 0.028$ Hz. The higher natural frequency is the first elastic frequency, $\omega_1 = 2.1$ Hz, which agrees with the semi–analytical solution. Also, the figure shows that the rigid body mode has a much higher amplitude than the elastic mode. In fact, the ratio between the amplitudes is about

$A(\Omega)/A(\omega_1) \cong 1000$, which means that the elastic motion is negligible from the point of view of structural dynamics.

Fig. 2.3 shows the tower's response to wave excitation at the times $t = 25, 50$, and $100\ s$. It is seen that the response is as a rigid body because the deflection y changes linearly with length x, as predicted in the semi–analytical solution.

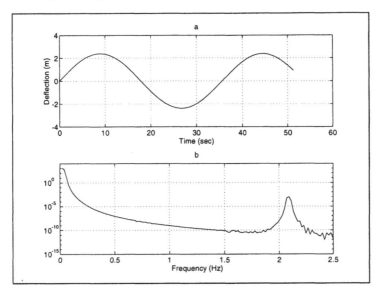

2.2 Free Vibration in the Time and Frequency Domains.

When the diameter of the tower is changed from $D = 15\ m$ to $D = 1.9\ m$, such that the first elastic natural frequency is $\omega_1 = 0.035\ Hz$, the response to the same wave excitation is totally different, as depicted in Fig. 2.4. The tower's response is no longer as a rigid body since its first elastic natural frequency is very close to not only the rigid body natural frequency, but also to the wave excitation frequency.

The conclusion from this analysis is that for most practical applications of an articulated tower, the rigid body model is representative of the lower order vibration. Such a simplified model is useful for exactly such preliminary conclusions.

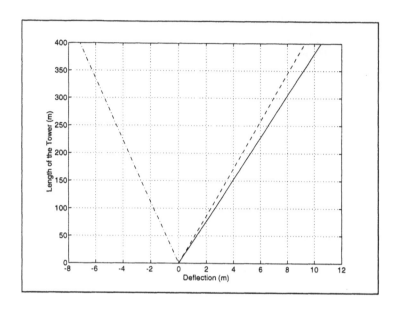

2.3 Response of the Tower to Wave Excitation at different times. Dash–dotted line: $t = 25s$, Solid line: $t = 50s$ and Dashed line: $t = 100s$.

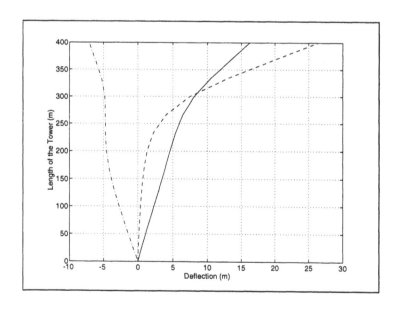

2.4 Response of the Flexible Tower to Wave Excitation at different times. Dash–dotted line: $t = 25s$, Solid line: $t = 50s$ and Dashed line: $t = 100s$.

Rigid Body Equations of Motion

We now proceed to derive the fully nonlinear rigid body equations of motion.

Problem Description
A schematic of the structure under study is shown in Fig. 2.5. It consists of a tower submerged in the ocean having a concentrated mass at the top and two degrees of freedom. These are θ about the z' axis and φ about the x'' axis. The tower is subjected to wave, current, wind and vortex shedding loads. As can be seen from Fig. 2.5, three coordinate systems are used in order to simplify the formulation of the equations of motion; one fixed (x,y,z), the second attached to the tower (x',y',z') in which the deflection angle θ is about z'. Finally, the third is rotating about x at φ (x'',y'',z''). All forces/moments, velocities, and accelerations are derived in the fixed coordinate system.

This problem has similarities to that of an inverted spherical pendulum with additional considerations;

- Buoyancy force is included.
- Drag force, proportional to the square of the relative velocity between the fluid and the tower, needs to be considered.
- Fluid inertia and added mass forces due to fluid and tower acceleration are part of the loading environment.
- Vortex shedding and wave slamming forces are considered.
- Current and wind forces are included.
- Earth angular velocity is included.

Lagrange's Equations
The equations of motion for large displacements are derived next using Lagrange's equation. Certain assumptions have been made and they are listed below.

- The tower has an infinite bending rigidity: $EI = \infty$.
- The hinge is the source of Coulomb friction.
- The end mass M is considered concentrated at the top. It is modeled as having the same diameter as the tower but different density.
- The tower diameter is much smaller than its length, $D \ll l$.
- The tower is a slender smooth structure with uniform cross section.
- The structure is at a stable static position due to the buoyancy force.
- Linear wave theory with random wave height is used.
- Wave forces are approximated via Morison's equation.

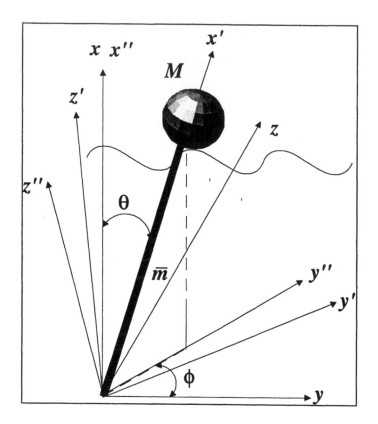

2.5 Model and Coordinate Systems.

The general form of Lagrange's equation is

$$\frac{d}{dt}\left(\frac{\partial K_E}{\partial \dot{q}_i}\right) - \frac{\partial K_E}{\partial q_i} + \frac{\partial P_E}{\partial q_i} + \frac{\partial D_E}{\partial \dot{q}_i} = Q_{q_i}, \qquad (2.55)$$

where K_E is the kinetic energy, P_E is the potential energy, D_E is the dissipative energy and Q_{q_i} is the generalized force related to the q_i generalized coordinate.

The model consists of two degrees of freedom, thus requiring two generalized coordinates, θ and φ. The generalized forces in the relevant directions are derived using the principle of virtual work. All external forces are resolved into the cartesian coordinates x, y, z, with unit vectors $\hat{x}, \hat{y}, \hat{z}$. Therefore, the general form for the force, assuming external force per unit length, has three components,

$$\mathbf{F}_e = F_x \hat{x} + F_y \hat{y} + F_z \hat{z}. \qquad (2.56)$$

From Fig. 2.6, the virtual work done by $\mathbf{F}_e$ due to a virtual displacement $\delta\theta$ is found to be

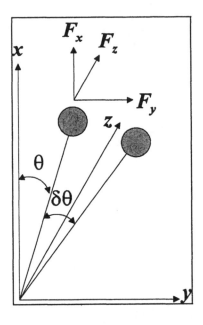

2.6 Generalized Force for θ.

$$
\begin{aligned}
F\theta\delta\theta &= F_x x'[\cos(\theta + \delta\theta) - \cos\theta] \\
&\quad +F_y x' \cos\varphi[\sin(\theta + \delta\theta) - \sin\theta] \\
&\quad +F_z x' \sin\varphi[\sin(\theta + \delta\theta) - \sin\theta],
\end{aligned} \tag{2.57}
$$

and using appropriate trigonometric identities

$$
\begin{aligned}
F\theta\delta\theta &= F_x x'[\cos\theta\cos\delta\theta - \sin\theta\sin\delta\theta - \cos\theta] \\
&\quad +F_y x' \cos\varphi[\sin\theta\cos\delta\theta + \cos\theta\sin\delta\theta - \sin\theta] \\
&\quad +F_z x' \sin\varphi[\sin\theta\cos\delta\theta + \cos\theta\sin\delta\theta - \sin\theta].
\end{aligned} \tag{2.58}
$$

By taking the limit of the expression above when virtual displacement approaches zero, $\delta\theta \to 0$, and making the replacement $x' = x/\cos\theta$, the generalized force per unit length for the θ coordinate is formed

$$
F\theta = -F_x x \tan\theta + F_y x \cos\varphi + F_z x \sin\varphi. \tag{2.59}
$$

From Fig. 2.7 the virtual work done by $\mathbf{F}_e$ due to a virtual displacement $\delta\varphi$ is derived,

$$
\begin{aligned}
F\varphi\delta\varphi &= F_y x' \sin\theta[\cos(\varphi + \delta\varphi) - \cos\varphi] \\
&\quad +F_z x' \sin\theta[\sin(\varphi + \delta\varphi) - \sin\varphi],
\end{aligned} \tag{2.60}
$$

and going through the same procedure described for $F\theta$, the generalized force per unit length is found to be

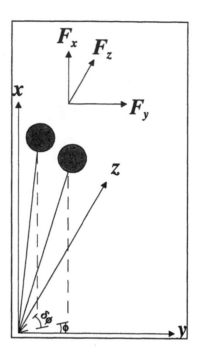

2.7 Generalized Force for φ.

$$F\varphi = -F_y x \tan\theta \sin\varphi + F_z x \tan\theta \cos\varphi. \tag{2.61}$$

Finally, the generalized moments are evaluated by integrating the moments due to $F\theta$ and $F\varphi$,

$$M\theta = \int_0^L (-F_x \tan\theta + F_y \cos\varphi + F_z \sin\varphi) x\,dx, \tag{2.62}$$

and

$$M\varphi = \int_0^L (-F_y \tan\theta \sin\varphi + F_z \tan\theta \cos\varphi) x\,dx, \tag{2.63}$$

where L is the projection in the x direction of the submerged part of the tower. L is a function of the deflection angle θ as follows :

$$L = \begin{cases} l\cos\theta & \text{if } d > l\cos\theta \\ d + \eta(y,t) & \text{if } d < l\cos\theta, \end{cases} \tag{2.64}$$

where d is the mean water level and $\eta(y,t)$ is the wave height elevation, to be defined later. Equations (2.62) and (2.63) can now be used to evaluate the moments due to the fluid.

Tower, Wave and Current Kinematics

To derive the equations of motion using Lagrange's equation, the kinetic, potential and dissipative energies, as well as the generalized forces, need to be evaluated. In this subsection, the tower's rectilinear and angular absolute velocities and accelerations are determined and expressed in the fixed coordinate system x,y,z.

Tower Kinematics

The tower is assumed to be oriented along a unit vector l with the following directional cosines (see Fig. 2.5)

$$l = \cos\theta\hat{x} + \sin\theta\cos\varphi\hat{y} + \sin\theta\sin\varphi\hat{z}, \tag{2.65}$$

so that the tower's radius vector $\mathbf{R}$ expressed in the x', y', z' coordinates is

$$\mathbf{R} = x'l. \tag{2.66}$$

Taking the first and second time derivatives of $\mathbf{R}$, and then replacing $x' = x/\cos\theta$, results in the following radius vector $\mathbf{R}$, velocity $\mathbf{V}$, and acceleration $\dot{\mathbf{V}}$,

$$\mathbf{R} = x\hat{x} + x\tan\theta\cos\varphi\hat{y} + x\tan\theta\sin\varphi\hat{z} \tag{2.67}$$

$$\mathbf{V} = -x\theta_t\tan\theta\hat{x} + x(\theta_t\cos\varphi - \varphi_t\tan\theta\sin\varphi)\hat{y} \tag{2.68}$$
$$+x(\theta_t\sin\varphi + \varphi_t\tan\theta\cos\varphi)\hat{z}$$

$$\dot{\mathbf{V}} = -x(\theta_{tt}\tan\theta + \theta_t^2)\hat{x} \tag{2.69}$$
$$+x[\theta_{tt}\cos\varphi - \varphi_{tt}\tan\theta\sin\varphi - (\theta_t^2 + \varphi_t^2)\tan\theta\cos\varphi - 2\theta_t\varphi_t\sin\varphi]\hat{y}$$
$$+ x[\theta_{tt}\sin\varphi + \varphi_{tt}\tan\theta\cos\varphi - (\theta_t^2 + \varphi_t^2)\tan\theta\sin\varphi + 2\theta_t\varphi_t\cos\varphi]\hat{z}.$$

Fig. 2.8 depicts the Earth coordinate system from which the tower's absolute angular velocity can be evaluated. The system X, Y, Z is an inertial coordinate system, X', Y', Z' is attached to Earth, where X' is normal to Earth, Y' is directed to east and Z' is directed to the north. The coordinate system x, y, z is attached to Earth with its origin at the tower's pivot. It is rotated with an angle β about the X' direction, and its y coordinate is in the direction of the ocean wave propagation. To simplify the calculations, the Earth angular velocity is expressed in the coordinate system x'', y'', z'', that rotates about x with the tower's angular rotation velocity φ_t.

The Earth angular velocity is

$$\mathbf{\Omega}_e = \Omega_e\hat{Z} = \Omega_e\sin\lambda\hat{X}' + \Omega_e\cos\lambda\hat{Z}', \tag{2.70}$$

where λ is the latitude angle. Transforming the Earth angular velocity to the rotating coordinate system yields

$$\mathbf{\Omega}_e = \Omega_e\sin\lambda\hat{x}'' + \Omega_e\cos\lambda\sin\beta\cos\varphi\hat{y}'' + \Omega_e\cos\lambda\cos\beta\sin\varphi\hat{z}''. \tag{2.71}$$

The tower's angular velocity relative to the Earth rotation is

$$\mathbf{\Omega}_t = \varphi_t\hat{x}'' + \theta_t\hat{z}''. \tag{2.72}$$

Finally, the absolute angular velocity of the tower is

$$\mathbf{\Omega}_T = \mathbf{\Omega}_t + \mathbf{\Omega}_e = (\Omega_e\sin\lambda + \varphi_t)\hat{x}'' + (\Omega_e\cos\lambda\sin\beta\cos\varphi)\hat{y}'' \tag{2.73}$$
$$+(\Omega_e\cos\lambda\cos\beta\sin\varphi + \theta_t)\hat{z}''.$$

Wave and Current Kinematics

In this study linear wave theory is assumed, implying that convective terms are negligible and only local velocities and accelerations are considered. Therefore, the wave vertical and horizontal velocities are (Wilson [4] pp. 84):

$$w_w = \frac{1}{2}H\omega\frac{\sinh kx}{\sinh kd}\sin(ky - \omega t)$$

$$u_w = \frac{1}{2}H\omega\frac{\cosh kx}{\sinh kd}\cos(ky - \omega t), \tag{2.74}$$

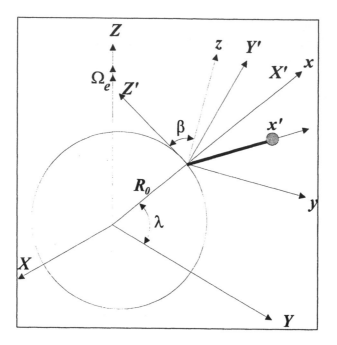

2.8 Earth and Tower Coordinate Systems.

and the respective accelerations:

$$\dot{w}_w = -\frac{1}{2}H\omega^2\frac{\sinh kx}{\sinh kd}\cos(ky - \omega t)$$

$$\dot{u}_w = \frac{1}{2}H\omega^2\frac{\cosh kx}{\sinh kd}\sin(ky - \omega t),\qquad (2.75)$$

where H is the significant wave height, ω the , k the, and d the mean water level, which are related by

$$\omega^2 = gk\tanh(kd).\qquad (2.76)$$

Without losing generality we assume that the waves propagate in the y direction so that the horizontal velocity u is in that direction, and w is in the x direction. We are aware of the fact that random waves are not unidirectional, but this consideration is beyond the scope of this study. If directional waves become important, one approach to a solution is to use the present results in a Monte Carlo simulation, where a directional probability distribution can be used for the wave direction.

Ocean current speed is calculated assuming that the current is made up of two different components (Isaacson (1988) [57]): the tidal component, U_c^t, and the wind–induced component U_c^w. If both components are known at the water surface, the vertical distribution of the current velocity $U_c(x)$ may be taken as

$$U_c(x) = U_c^t\left(\frac{x}{d}\right)^{\frac{1}{7}} + U_c^w\left(\frac{x}{d}\right).\qquad (2.77)$$

The tidal current U_c^t at the surface can be obtained directly from the tide table, and the wind–driven current U_c^w at the surface is generally taken as 1 to 5 % of the mean wind speed at $10\ m$ above the surface.

When current and waves coexist, the combined flow field should be used to determine the wave loads. The influence of an assumed uniform current on the wave field is treated by applying wave theory in a reference frame which is fixed relative to the current. For a current of magnitude U_c propagating in a direction α relative to the direction of the wave propagation, the wave velocity, $c_0 = \omega_0/k$ for no current, is modified and becomes

$$c = c_0 + U_c \cos \alpha. \tag{2.78}$$

The velocities then used to determine the wave loads are the vectorial sum of the wave and current velocities

$$w = w_w \tag{2.79}$$

$$u = u_w + U_c \cos \alpha \tag{2.80}$$

$$v = U_c \sin \alpha, \tag{2.81}$$

where u, v and w are the total velocities in x, y and z directions, respectively.

To consider geometric nonlinearities, the velocities and accelerations are evaluated at the instantaneous position of the tower. Replacing $y = x \tan \theta \cos \varphi$ in the velocity and acceleration expressions (2.74), (2.75) yields velocities

$$w = \frac{1}{2}H\omega\frac{\sinh kx}{\sinh kd}\sin(kx\tan\theta\cos\varphi - \omega t) \tag{2.82}$$

$$u = \frac{1}{2}H\omega\frac{\cosh kx}{\sinh kd}\cos(kx\tan\theta\cos\varphi - \omega t) + U_c \cos\alpha \tag{2.83}$$

$$v = U_c \sin\alpha, \tag{2.84}$$

and accelerations

$$\dot{w} = -\frac{1}{2}H\omega^2\frac{\sinh kx}{\sinh kd}\cos(kx\tan\theta\cos\varphi - \omega t) \tag{2.85}$$

$$\dot{u} = \frac{1}{2}H\omega^2\frac{\cosh kx}{\sinh kd}\sin(kx\tan\theta\cos\varphi - \omega t) \tag{2.86}$$

$$\dot{v} = 0. \tag{2.87}$$

The influence of current on the significant wave height depends on the manner in which the waves propagate onto the current field. An approximation to the significant wave height in the presence of current is given by Isaacson (1988) [57] as

$$H = H_0\sqrt{\frac{2}{\gamma + \gamma^2}}, \tag{2.88}$$

where H_0 and H are the significant wave heights in the absence and presence of current, respectively, and γ is defined as

$$\gamma = \sqrt{1 + \frac{4U_c}{c_0}\cos\alpha} \quad \text{for} \quad \frac{4U_c}{c_0}\cos\alpha > -1. \tag{2.89}$$

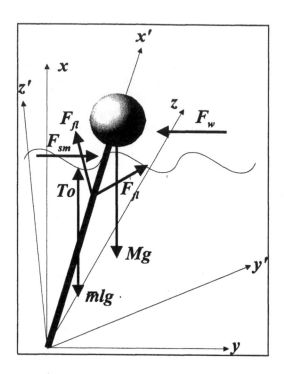

2.9 External Forces Acting on the Tower.

Fluid Forces and Moments Acting on the Tower
    Fig. 2.9 depicts the external forces acting on the tower,

- T_0 is the buoyancy force.
- F_{fl} are the vertical and horizontal fluid forces due to drag, inertia, added mass and vortex shedding.
- Mg and $\bar{m}lg$ are the forces due to gravity.
- F_w is the wind force.
- F_{sm} is the wave slamming load.

    Next explicit expressions for these forces and moments are described and developed .

    Buoyancy Moment: The force that provides the restoring moment is assumed to be generated by the whole length of the tower,

$$M\theta^b = T_0 l_b. \tag{2.90}$$

T_0 is the time–dependent resultant buoyancy force,

$$T_0 = \rho g V_0 = \rho g \pi \frac{D^2}{4} L_s, \tag{2.91}$$

and l_b is its moment arm, V_0 is the volume of the submerged part of the tower, ρ is the fluid density and L_s, which is the length of the submerged part of the tower, equals

$$L_s = \begin{cases} \frac{d+\eta(y,t)}{\cos\theta} & \text{if } d + \eta(y,t) < l\cos\theta \\ l & \text{if } d + \eta(y,t) > l\cos\theta, \end{cases} \qquad (2.92)$$

where $\eta(y,t)$ is the wave height elevation evaluated at the instantaneous position of the tower with $y = d\tan\theta\cos\varphi$ at $x = d$,

$$\eta(\theta,\varphi,t) = \frac{1}{2}H\cos(kd\tan\theta\cos\varphi - \omega t + \varepsilon), \qquad (2.93)$$

and ε is a random phase.

The buoyant force acts at the center of mass of the submerged part of the tower. If we consider the tower to be of circular cylindrical cross–section, then the center of mass in the x', y', z' coordinates is

$$l_b^{y'} = \frac{D^2}{16L_s}\tan\theta \qquad (2.94)$$

$$l_b^{x'} = \frac{1}{2}L_s + \frac{D^2}{32L_s}\tan^2\theta. \qquad (2.95)$$

Transforming to x, y, z coordinates, we find the moment arm l_b to be

$$l_b = l_b^{x'}\sin\theta + l_b^{y'}\cos\theta \qquad (2.96)$$

or

$$l_b = \left(\frac{1}{2}L_s + \frac{D^2}{32L_s}\tan^2\theta\right)\sin\theta + \frac{D^2}{16L_s}\tan\theta\cos\theta, \qquad (2.97)$$

and finally the buoyancy generalized moment is

$$M\theta^b = \rho g\pi\frac{D^2}{4}\sin\theta\left[\frac{D^2}{32}\left(2+\tan^2\theta\right) + \frac{1}{2}\left(\frac{d+\eta(y,t)}{\cos\theta}\right)^2\right]. \qquad (2.98)$$

Drag and Inertia Forces – Morison's Equation: The drag and inertia forces per unit length are approximated by Morison's equation (Morison et al. [2]), in which the following is assumed; the body is cylindrical, the wave slope and the associated pressure gradient are constant across the diameter of the body and the wave scattering is negligible. The drag force is proportional to the square of the relative velocity between the fluid and the tower, and the inertia force is proportional to the fluid acceleration,

$$\mathbf{F}_{fl} = C_D\rho\frac{D}{2}\mid\mathbf{V_{rel}}\mid\mathbf{V_{rel}} + C_M\rho\pi\frac{D^2}{4}\dot{\mathbf{U}}_w, \qquad (2.99)$$

where $\mathbf{F}_{fl}$ is the fluid force per unit length normal to the tower. $\mathbf{V_{rel}}$ is the vector of the relative velocity between the fluid and the tower in a direction normal to the tower, and $\dot{\mathbf{U}}_w$ is the fluid acceleration normal to the tower. The added mass effect that can be included in the mass term of the Morison equation is considered separately in the next section. C_D and C_M are the drag and inertia coefficients, respectively. For small angles, say, $\theta < 0.15\ rad$, the relative velocity $\mathbf{V_{rel}}$ can be taken in the direction of the wave propagation, resulting in a good approximation. The relative velocity and fluid acceleration normal to the tower can be decomposed to their components as follows,

$$\begin{bmatrix} V_{rel}^x \\ V_{rel}^y \\ V_{rel}^z \end{bmatrix} = 1 \times (\mathbf{U_w} - \mathbf{V}) \times 1$$

and

$$\begin{bmatrix} \dot{U}_w^x \\ \dot{U}_w^y \\ \dot{U}_w^z \end{bmatrix} = 1 \times \dot{\mathbf{U}}_\mathbf{w} \times 1. \tag{2.100}$$

Using Morison's equation (2.99), the tower velocity equation (2.68), and fluid velocity and acceleration equations (2.82)–(2.84) and (2.85)–(2.87), the fluid force components are:

the drag force

$$\begin{bmatrix} F_{xD}^{fl} \\ F_{yD}^{fl} \\ F_{zD}^{fl} \end{bmatrix} = C_D \rho \frac{D}{2} \sqrt{(V_{rel}^x)^2 + (V_{rel}^y)^2 + (V_{rel}^z)^2} \begin{bmatrix} V_{rel}^x \\ V_{rel}^y \\ V_{rel}^z \end{bmatrix}, \tag{2.101}$$

and the inertia force

$$\begin{bmatrix} F_{xI}^{fl} \\ F_{yI}^{fl} \end{bmatrix} = C_M \rho \pi \frac{D^2}{4} \begin{bmatrix} \dot{w} \\ \dot{u} \end{bmatrix}. \tag{2.102}$$

Added Mass Moment: The fluid added mass force per unit length $\mathbf{F_{ad}}$ is

$$\mathbf{F_{ad}} = C_A \rho \pi \frac{D^2}{4} \dot{\mathbf{V}}, \tag{2.103}$$

where $C_A = C_M - 1$ is the added mass coefficient. Substituting equation (2.69) for the tower acceleration into equation (2.103) leads to the expressions for the forces in the $x, y,$ and z directions,

$$F_x^{ad} = -C_A \rho \pi \frac{D^2}{4} x (\theta_{tt} \tan \theta + \theta_t^2) \tag{2.104}$$

$$F_y^{ad} = C_A \rho \pi \frac{D^2}{4} x [\theta_{tt} \cos \varphi - \varphi_{tt} \tan \theta \sin \varphi \tag{2.105}$$
$$- (\theta_t^2 + \varphi_t^2) \tan \theta \cos \varphi - 2\theta_t \varphi_t \sin \varphi]$$

$$F_z^{ad} = C_A \rho \pi \frac{D^2}{4} x [\theta_{tt} \sin \varphi + \varphi_{tt} \tan \theta \cos \varphi \tag{2.106}$$
$$- (\theta_t^2 + \varphi_t^2) \tan \theta \sin \varphi + 2\theta_t \varphi_t \cos \varphi].$$

Substituting the added mass forces, equations (2.104)–(2.106), into the generalized moment equations (2.62) and (2.63), and integrating, results in the generalized moments due to fluid added mass,

$$M\theta^{ad} = \frac{1}{12} C_A \rho \pi \frac{D^2}{4} L^3 \left(\theta_{tt} \left(1 + \tan^2 \theta \right) + \varphi_t^2 \tan \theta \right) \tag{2.107}$$

$$M\varphi^{ad} = \frac{1}{12} C_A \rho \pi \frac{D^2}{4} L^3 \left(\varphi_{tt} \tan^2 \theta + 2\theta_t \varphi_t \tan \theta \right). \tag{2.108}$$

Vortex Shedding Force: The lift force $\mathbf{F_L}$ due to vortex shedding is acting in a direction normal to the wave velocity vector and normal to the tower. Different models of the vortex shedding forces exist in the literature; see especially Billah (1989) [58] . A simple model given in a paper by Dong (1991) [59] is used here,

$$\mathbf{F_L} = C_L \rho \frac{D}{2} \cos \omega_s t \, |1 \times (\mathbf{U_T} - \mathbf{V})| \, (1 \times (\mathbf{U_T} - \mathbf{V})), \qquad (2.109)$$

where the vector of the maximum fluid velocity along the tower $\mathbf{U_T}$ is

$$\begin{bmatrix} U_T^x \\ U_T^y \\ U_T^z \end{bmatrix} = \begin{bmatrix} \frac{1}{2} H \omega \frac{\sinh kx}{\sinh kd} \\ \frac{1}{2} H \omega \frac{\cosh kx}{\sinh kd} + U_c \cos \alpha \\ U_c \sin \alpha \end{bmatrix} . \qquad (2.110)$$

C_L is the lift coefficient, and ω_s is the vortex shedding frequency (Issacson (1988) [57]), given by

$$\omega_s = \begin{cases} 2\omega, & H \neq 0 \\ \frac{St U_c}{D}, & H = 0, \end{cases} \qquad (2.111)$$

where St is the *Strouhal number*, which varies slightly with *Reynolds number*, *Re*, but is roughly 0.2 over a wide practical range in *Re*.

It is understood here that more sophisticated models of the vortex–structure interaction are possible and desirable.

Wave Slamming Force: The parts of a structural member that are about the mean water level are exposed to an impulsive force caused by wave slamming. The slamming force per unit length has a form similar to the drag force (see Chakrabarti (1990) pp. 142–143 [60]),

$$\mathbf{F_s} = C_S \rho \frac{D}{2} \, |1 \times \mathbf{U_S} \times 1| \, (1 \times \mathbf{U_S} \times 1), \qquad (2.112)$$

where C_S the wave slamming coefficient with a theoretical value of π, but a typical mean value may be taken as 3.5 even though considerable scatter in this coefficient has been found in laboratory experiments. $\mathbf{U_S}$ is the relative velocity between the waves and the tower at the zone of impact, where

$$\begin{bmatrix} U_S^x \\ U_S^y \\ U_S^z \end{bmatrix} = \begin{bmatrix} \frac{1}{2} H \omega \frac{\sinh kx}{\sinh kd} + x\theta_t \tan\theta \\ \frac{1}{2} H \omega \frac{\cosh kx}{\sinh kd} + U_c \cos \alpha - x(\theta_t \cos\varphi - \varphi_t \tan\theta \sin\varphi) \\ U_c \sin \alpha - x(\theta_t \sin\varphi + \varphi_t \tan\theta \cos\varphi) \end{bmatrix} . \qquad (2.113)$$

This force is assumed to be a periodic impulse with the period of the wave, and in duration on the order of milliseconds, as described in Faltinsen, pp. 282–285 [8] . The wave slamming moments are evaluated by integrating the slamming force along the part of the tower onto which the waves are slamming. Since the exact length is very complicated to determine, an approximate length equal to the distance from the mean water level d to the wave height $d + \frac{1}{2} H$ is used, which was suggested by Chakrabarti, pp. 142–143 [60] . Hence,

$$M\theta^s = \int_d^{d+\frac{1}{2}H} \left(-F_x^{sm} \tan\theta + F_y^{sm} \cos\varphi + F_z^{sm} \sin\varphi \right) x \, dx \qquad (2.114)$$

$$M\varphi^s = \int_d^{d+\frac{1}{2}H} \left(-F_y^{sm} \tan\theta \sin\varphi + F_z^{sm} \tan\theta \cos\varphi \right) x \, dx, \qquad (2.115)$$

where F_x^{sm}, F_y^{sm}, and F_z^{sm} are the wave slamming forces in the $x, y,$ and z directions.

Total Fluid Moment: The moments due to the fluid forces (drag, inertia, and lift) are evaluated by substituting the sum of all fluid forces, as defined by the following set of equations (2.116)–(2.118),

$$F_x^{fl} = F_{xD}^{fl} + F_{xI}^{fl} + F_{xL}^{fl} \qquad (2.116)$$

$$F_y^{fl} = F_{yD}^{fl} + F_{yI}^{fl} + F_{yL}^{fl} \qquad (2.117)$$

$$F_z^{fl} = F_{zD}^{fl} + F_{zI}^{fl} + F_{zL}^{fl}, \qquad (2.118)$$

into the general fluid moments, equations (2.62) and (2.63). Therefore the fluid moments $M\theta^{fl}$ and $M\varphi^{fl}$ are given by

$$M\theta^{fl} = \int_0^L \left(-F_x^{fl} \tan\theta + F_y^{fl} \cos\varphi + F_z^{fl} \sin\varphi \right) x\, dx \qquad (2.119)$$

$$M\varphi^{fl} = \int_0^L \left(-F_y^{fl} \tan\theta \sin\varphi + F_z^{fl} \tan\theta \cos\varphi \right) x\, dx, \qquad (2.120)$$

where the superscript notation fl has been added so that it is possible to keep better track of all the component moments.

Wind Load: Wind loads, drag and lift, are derived in a way similar to current forces. Both the drag and lift force expressions used are similar to those of the fluid. The drag force is

$$\mathbf{F_D^w} = C_{Da}\rho_a \frac{D}{2} \left[|1 \times (\mathbf{u}_w - \mathbf{V}) \times 1| \, (1 \times (\mathbf{u}_w - \mathbf{V}) \times 1) \right],$$

where C_{Da} is the air drag coefficient, ρ_a is the air density and $\mathbf{u}_w$ is the wind velocity, which is assumed to propagate in an arbitrary direction,

$$\mathbf{u}_{wind} = u_{wind} \cos\nu \hat{y} + u_{wind} \sin\nu \hat{z}, \qquad (2.121)$$

where ν is the angle between the propagating wind and the y axis. The lift force due to vortex shedding is

$$\mathbf{F_L^w} = C_{La}\rho_a \frac{D}{2} \cos\omega_{Lw} t \, |1 \times (\mathbf{u}_{wind} - \mathbf{V})| \, (1 \times (\mathbf{u}_{wind} - \mathbf{V})), \qquad (2.122)$$

where C_{La} is the air lift coefficient. The vortex shedding frequency ω_{Lw} is taken to be

$$\omega_{Lw} = \frac{u_w St}{D}. \qquad (2.123)$$

The moments due to wind loads, $M\theta^w$ and $M\varphi^w$, are found in using equations (2.119) and (2.120), but these are integrated along the exposed part of the tower,

$$M\theta^w = \int_{d+\eta}^{l\cos\theta} \left(-F_x^w \tan\theta + F_y^w \cos\varphi + F_z^w \sin\varphi \right) x\, dx \qquad (2.124)$$

$$M\varphi^w = \int_{d+\eta}^{l\cos\theta} \left(-F_y^w \tan\theta \sin\varphi + F_z^w \tan\theta \cos\varphi \right) x\, dx, \qquad (2.125)$$

where F_x^w, F_y^w, and F_z^w, are the wind forces in the $x, y,$ and z directions, due to drag and vortex shedding. If the tower is fully submerged, the wind moments are zero, $M\theta^w = M\varphi^w = 0$.

Friction Moment: Dissipation in the tower hinge is assumed to be modeled as Coulomb friction. In this section the friction/damping moment is evaluated. The friction force is equal to the product of the normal force at the hinge N and the coefficient of friction μ, which is assumed to be independent of the velocity once the motion is initiated. Since the sign of the damping force is always opposite that of the velocity, the differential equation of motion for each sign is valid only for half cycle intervals. The friction force is modelled as

$$F\theta^{fr} = N\mu[sgn(\theta_t)] \tag{2.126}$$
$$F\varphi^{fr} = N\mu[sgn(\varphi_t)], \tag{2.127}$$

with the normal force given by

$$N = \sum F_x \cos\theta + \sum F_y \cos\varphi \sin\theta + \sum F_z \sin\varphi \sin\theta, \tag{2.128}$$

and $\sum F_x$, $\sum F_y$ and $\sum F_z$ are the sum of the forces due to gravity, buoyancy and tower acceleration in the x, y and z directions, respectively. The fluid drag, inertia and vortex shedding forces do not influence the friction force since they act perpendicularly to the tower. Thus,

$$\sum F_x = T_0 - F_g + F_x^{ac} \tag{2.129}$$
$$\sum F_y = F_y^{ac} \tag{2.130}$$
$$\sum F_z = F_z^{ac}, \tag{2.131}$$

where T_0 is the buoyancy force given in equation (2.91), F_g is the gravitational force,

$$F_g = (\bar{m}l + M)g, \tag{2.132}$$

and the forces due to the tower acceleration are, using equation (2.69),

$$F_x^{ac} = \left[\frac{1}{8}C_A\rho\pi D^2 L^2 + \frac{1}{2}\left(\frac{1}{2}\bar{m}l + M\right)\bar{l}\right]\frac{1}{\cos\theta}\left(\theta_{tt}\tan\theta + \theta_t^2\right) \tag{2.133}$$

$$F_y^{ac} = \left[\frac{1}{8}C_A\rho\pi D^2 L^2 + \frac{1}{2}\left(\frac{1}{2}\bar{m}l + M\right)\bar{l}\right]\frac{1}{\cos\theta} \cdot \tag{2.134}$$
$$\cdot \left(-\theta_{tt}\cos\varphi + \varphi_{tt}\tan\theta\sin\varphi + (\theta_t^2 + \varphi_t^2)\tan\theta\cos\varphi + 2\theta_t\varphi_t\sin\varphi\right)$$

$$F_z^{ac} = \left[\frac{1}{8}C_A\rho\pi D^2 L^2 + \frac{1}{2}\left(\frac{1}{2}\bar{m}l + M\right)\bar{l}\right]\frac{1}{\cos\theta} \cdot \tag{2.135}$$
$$\cdot \left(-\theta_{tt}\sin\varphi - \varphi_{tt}\tan\theta\cos\varphi + (\theta_t^2 + \varphi_t^2)\tan\theta\sin\varphi - 2\theta_t\varphi_t\cos\varphi\right),$$

where $\bar{l}$ is the projection of the tower's length l in the x direction, $\bar{l} = l\cos\theta$. Assuming a hinge radius R_h, and rearranging, the friction moments are found

$$M\theta^{fr} = R_h N\mu[sgn(\theta_t)] \tag{2.136}$$
$$M\varphi^{fr} = R_h \sin\theta N\mu[sgn(\varphi_t)], \tag{2.137}$$

where N, the normal force, is explicitly

$$N = \left[\frac{1}{8}C_A\rho\pi D^2\frac{L^2}{\cos^2\theta} + \frac{1}{2}\left(\frac{1}{2}\bar{m}l + M\right)l\right](\theta_t^2 + \frac{1}{2}\varphi_t^2)$$

$$- \left[\frac{1}{8} C_A \rho \pi D^2 L^2 + \frac{1}{2} \left(\frac{1}{2} \bar{m}l + M \right) l \cos 2\theta \right] \frac{1}{2} \varphi_t^2 \qquad (2.138)$$

$$+ (T_0 - F_g) \cos \theta.$$

The only remaining inertial forces are due to centrifugal acceleration, which are along the tower, that is, $l\theta_t^2$ and $l\varphi_t^2$.

Dynamic moments: The dynamic moments, $M\theta^{dy}$ and $M\varphi^{dy}$, which would appear on the left hand side of Lagrange's equation (2.55), are found using expressions for the , potential, and dissipative energies,

$$K_E = \frac{1}{2} (I_{x''} \Omega_{Tx}^2 + I_{y''} \Omega_{Ty}^2 + I_{z''} \Omega_{Tz}^2) \qquad (2.139)$$

$$P_E = (\frac{1}{2} \bar{m}l + M) gl \cos \theta \qquad (2.140)$$

$$D_E = \frac{1}{2} C \left(\Omega_{tx}^2 + \Omega_{ty}^2 + \Omega_{tz}^2 \right), \qquad (2.141)$$

where C is the structural damping constant and $I_{x''}$, $I_{y''}$, and $I_{z''}$ are the moments of inertia of the tower, given by

$$I_{x''} = (\frac{1}{3} \bar{m}l + M) l^2 \sin^2 \theta + \frac{1}{2} (\bar{m}l + M) \frac{D^2}{4} \cos^2 \theta \qquad (2.142)$$

$$I_{y''} = (\frac{1}{3} \bar{m}l + M) l^2 \cos^2 \theta + \frac{1}{2} (\bar{m}l + M) \frac{D^2}{4} \sin^2 \theta \qquad (2.143)$$

$$I_{z''} = \left(\frac{1}{3} \bar{m}l + M \right) l^2, \qquad (2.144)$$

where the Ω term are defined in equations (2.72) and (2.73). Substitute these and (2.142)–(2.144) into (2.139) and (2.141), leading to expressions for the kinetic and dissipative energies,

$$K_E = \frac{1}{2} \left((\frac{1}{3} \bar{m}l + M) l^2 \sin^2 \theta + \frac{1}{2} (\bar{m}l + M) \frac{D^2}{4} \cos^2 \theta \right) \cdot \qquad (2.145)$$

$$(\Omega_e \sin \lambda + \varphi_t)^2$$

$$+ \frac{1}{2} \left((\frac{1}{3} \bar{m}l + M) l^2 \cos^2 \theta + \frac{1}{2} (\bar{m}l + M) \frac{D^2}{4} \sin^2 \theta \right) \cdot$$

$$(\Omega_e \cos \lambda \sin \beta \cos \varphi)^2$$

$$+ \frac{1}{2} (\frac{1}{3} \bar{m}l + M) l^2 (\Omega_e \cos \lambda \cos \beta \sin \varphi + \theta_t)^2$$

$$D_E = \frac{1}{2} C \left(\varphi_t^2 + \theta_t^2 \right). \qquad (2.146)$$

Substituting the kinetic, potential and dissipative energies into Lagrange's equation (2.55) leads, after some mathematical manipulations and rearranging, to the dynamic moments $M\theta^{dy}$ and $M\varphi^{dy}$,

$$M\theta^{dy} = (\frac{1}{3} \bar{m}l + M) l^2 \theta_{tt} + C\theta_t - \left(\frac{1}{2} \bar{m}l + M \right) gl \sin \theta \qquad (2.147)$$

$$+ \left[\frac{l^2}{2} (\frac{1}{3} \bar{m}l + M) - \frac{D^2}{16} (\bar{m}l + M) \right] \cdot$$

$$\left[(\Omega_e \sin \lambda + \varphi_t)^2 + (\Omega_e \cos \lambda \sin \beta \cos \varphi)^2\right] \sin 2\theta$$

$$M\varphi^{dy} = \left(((\frac{1}{3}\bar{m}l + M)l^2 \sin^2 \theta + \frac{1}{8}(\bar{m}l + M)D^2 \cos^2 \theta\right) \varphi_{tt} + C\varphi_t$$

$$+ \left[\frac{l^2}{2}(\frac{1}{3}\bar{m}l + M) - \frac{D^2}{8}(\bar{m}l + M)\right] \sin 2\theta \left[\Omega_e \sin \lambda + \varphi_t\right]\theta_t$$

$$-\frac{1}{2}I_{y''}(\Omega_e \cos \lambda \sin \beta)^2 \sin 2\varphi$$

$$+I_{z''}\left(\frac{1}{2}\Omega_e \cos \lambda \cos \beta \sin 2\varphi + \theta_t \Omega_e \cos \lambda \cos \beta \cos \varphi\right). \tag{2.148}$$

Governing Equations of Motion

The governing nonlinear differential equations of motion are found by equating the dynamic moments to the applied external moments,

$$M\theta^{dy} = M\theta^{ap} \tag{2.149}$$

$$M\varphi^{dy} = M\varphi^{ap}. \tag{2.150}$$

The applied moments are found by adding all the moment equations, which results in

$$M\theta^{ap} = M\theta^b + M\theta^{fl} + M\theta^w + M\theta^s - M\theta^{fr} - M\theta^{ad} \tag{2.151}$$

$$M\varphi^{ap} = M\varphi^b + M\varphi^{fl} + M\varphi^w + M\varphi^s - M\varphi^{fr} - M\varphi^{ad}. \tag{2.152}$$

Substituting equations (2.147), (2.148) and (2.151), (2.152) into (2.149) and (2.150), and rearranging terms, leads to the governing nonlinear differential equations of motion for the tower;

$$J\theta^{eff}\theta_{tt} + I_g \left[(\Omega_e \sin \lambda + \varphi_t)^2 + (\Omega_e \cos \lambda \sin \beta \cos \varphi)^2\right] \sin 2\theta +$$

$$C\theta_t = M\theta^{fl} + M\theta^w + M\theta^s - M\theta^{fr} - M\theta^{gb} \tag{2.153}$$

$$J\varphi^{eff}\varphi_{tt} + I_g \left[\Omega_e \sin \lambda + \varphi_t\right]\theta_t \sin 2\theta + \frac{1}{2}I_{y''}(\Omega_e \cos \lambda \sin \beta)^2 \sin 2\varphi$$

$$+I_{z''}\left(\frac{1}{2}\Omega_e \cos \lambda \cos \beta \sin 2\varphi + \theta_t \Omega_e \cos \lambda \cos \beta \cos \varphi\right) +$$

$$C\varphi_t = M\varphi^{fl} + M\varphi^w + M\varphi^s - M\varphi^{fr}, \tag{2.154}$$

where some of the applied moments have been moved to the left hand sides of the governing equation, and where $J\theta^{eff}$ and $J\varphi^{eff}$ are the effective position dependent moments of inertia,

$$J\theta^{eff} = \left(\frac{1}{3}\bar{m}l + M\right)l^2 + \frac{1}{12}C_A\rho\pi D^2 L^3(1 + \tan^2 \theta) \tag{2.155}$$

$$J\varphi^{eff} = \left(\frac{1}{3}\bar{m}l + M\right)l^2 \sin^2 \theta + \frac{1}{8}(\bar{m}l + M)D^2 \cos^2 \theta$$

$$+\frac{1}{12}C_A\rho\pi D^2 L^3 \tan^2\theta. \tag{2.156}$$

I_g is a coefficient depending on the system parameters,

$$I_g = \left(\frac{l^2}{2}(\frac{1}{3}\bar{m}l + M) - \frac{D^2}{8}(\bar{m}l + M)\right) \tag{2.157}$$

and $M\theta^{gb}$ is the moment due to gravity and buoyancy,

$$M\theta^{gb} = \rho g\pi\frac{D^2}{4}\left[\frac{D^2}{32}\tan^2\theta(2\cos\theta + \sin\theta) + \frac{1}{2}\left(\frac{d + \eta(y,t)}{\cos\theta}\right)^2 \sin\theta\right]$$

$$-\left(\frac{1}{2}\bar{m}l + M\right)gl\sin\theta. \tag{2.158}$$

The equations can be simplified by neglecting the Coriolis acceleration terms due to Earth rotation. The resulting equations of motion,

$$J\theta^{eff}\theta_{tt} + C\theta_t + I_g\varphi_t^2 + M\theta^{gb} = M\theta^{fl} + M\theta^w + M\theta^s - M\theta^{fr} \tag{2.159}$$
$$J\varphi^{eff}\varphi_{tt} + C\varphi_t + I_g\varphi_t\theta_t = M\varphi^{fl} + M\varphi^w + M\varphi^s - M\varphi^{fr} \tag{2.160}$$

are similar to those derived by Kirk and Jain [35] , where appropriate simplifications are made of the above expressions for $J\theta^{eff}$ and $J\varphi^{eff}$.

Numerical Solution

In this section, the response of the two degree of freedom tower is investigated. First the response to deterministic parameter values is investigated, and then the average response and the standard deviation to uniformly distributed random parameters is evaluated. The physical parameters used in the simulations are presented in Tables 2.1 and 2.2, in consistent metric units,

l	D	M	$\bar{m}$	μ	R_h
350	15	25×10^5	20×10^3	$0.1 - 0.4$	1.5

2.1 Tower Properties

d	C_D	C_M	C_L	ρ	H	ω	U_c
400	0.6 to 2	1.0 to 2	0.8 to 1.2	1025	0 to 10	0.03 to 1	2

2.2 Fluid Properties

Deterministic Response

The equations of motion (2.153) and (2.154) are solved for the following cases:

- Free vibration with damping (drag, viscous and friction)
- Equilibrium position of the tower is determined in the presence of wind and/or current
- Response to wave, wind and current
- Superharmonic, harmonic and subharmonic resonances
- Response due to wave slamming and Coriolis acceleration
- Chaotic response.

Free Vibration: To find the fundamental frequency, the wave height, current, and wind velocities are set to zero. The initial condition is set to $\dot{\theta}_0 = 0.01$ *rad/s*. Figures 2.10, 2.11 and 2.12 depict the free vibration of the tower in the presence of different damping mechanisms; structural viscous damping, Coulomb friction and drag force, respectively. As can be seen from the figures, the natural frequency is $\omega_n = 0.028$ *Hz*. A typical decay for each damping mechanism is clearly seen: hyperbolic decay, proportional to $\theta a_1/t$, for drag damping in Fig. 2.12, linear decay, proportional to $(\theta - a_2 t)$, for Coulomb damping in Fig. 2.11 and exponential decay, $\theta e^{(-a_3 t)}$, for viscous damping in Fig. 2.10. The parameters a_1, a_2 and a_3 are the decay constants for each damping mechanism and are functions of the drag C_D, friction μ, and viscous damping ς coefficients respectively. The responses for the first two damping mechanisms consist of the fundamental frequency and its odd multipliers, as can be seen from the frequency domain figures. The reason for the odd multipliers is the fact that the drag and the Coulomb friction forces are nonlinear and antisymmetric. On the other hand, the response for viscous damping is linear and therefore only the fundamental frequency is seen.

Equilibrium Position: As was shown in the solution for the single degree of freedom model, see Bar-Avi and Benaroya [33], the equilibrium position depends on the current velocity and the drag coefficient. In this section, the equilibrium position due to current and wind is found. Also the effect of vortex shedding force on the response is investigated. The wave height is $H = 0$, the current velocity is $U_c = 1$ *m/s* and the wind velocity is $u_w = 20$ *m/s*. Fig. 2.13 shows the response due to current and wind propagating in the y direction, $\alpha = 0, \nu = 0$, respectively and with lift coefficient $C_L = 0$. After the transient decays, the tower remains at a constant angle, $\theta = 0.006$ *rad*, where all forces are in equilibrium.

When the current is in the z direction, with $C_L = 0$, the equilibrium position is at $y \approx 0.4$ *m* and $z \approx 2.0$ *m*, as can be seen from Fig. 2.14. The reason is that the total force due to wind and current is a vector summation.

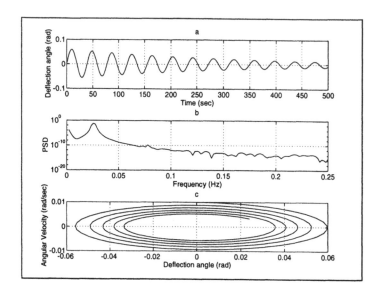

2.10 Damped Free–Vibration - Structural Viscous Damping, $\zeta = 0.02$.

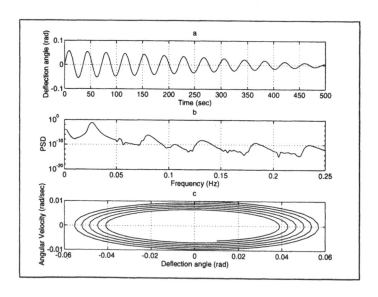

2.11 Damped Free–Vibration - Coulomb Friction, $\mu = 0.1$.

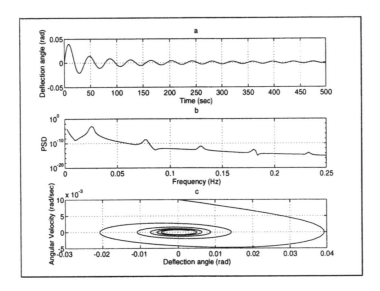

2.12 Damped Free–Vibration - Drag Force, $C_D = 1.2$.

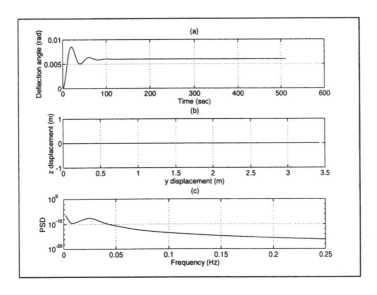

2.13 Equilibrium Position in the Presence of Current and Wind, $C_L = 0, \alpha = 0, \nu = 0$.
(a) θ Time Domain, (b) Tower's Top Motion, (c) θ Frequency Domain.

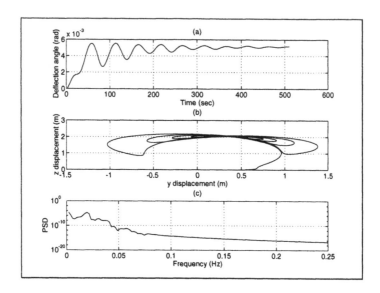

2.14 Equilibrium Position in the Presence of Current and Wind, $C_L = 0$, $\alpha = 90^0$, $\nu = 0$. (a) θ Time Domain, (b) Tower's Top Motion, (c) θ Frequency Domain.

The simplified expression for the equilibrium position is derived by linearizing equations (2.159) and (2.160) and solving them in the presence of wind and current,

$$\theta_{eq} = \frac{1}{2} \frac{\sqrt{(F_{U_c})^2 + (F_{U_w})^2 + 2F_{U_c} F_{U_w} \cos(\alpha - \nu)}}{\rho g \pi D^2 d^2 - 2gl(\bar{m}l + 2M)} \tag{2.161}$$

$$\varphi_{eq} = \tan^{-1}\left(\frac{F_{U_c} \sin\alpha + F_{U_w} \sin\nu}{F_{U_c} \cos\alpha + F_{U_w} \cos\nu} \right), \tag{2.162}$$

where

$$F_{U_c} = C_D \rho D d^2 U_c^2 \tag{2.163}$$

$$F_{U_w} = C_{Da} \rho_a D(l-d)^2 u_w^2. \tag{2.164}$$

Setting the lift coefficient $C_L = 1$ changes the response of the tower to current and wind as can be seen from the next two figures. Fig. 2.15 depicts the response when both the wind and current are in the y direction. The tower's top oscillates with an amplitude of about $y = 2.4\ m$ in the z direction. The oscillations are in the range of vortex shedding frequencies due to current, $\omega_s = 0.002\ Hz$, and due to wind, $\omega_{Lw} = 0.04$ Hz, as can be seen from Fig. 2.15 (c). Note in Fig. 2.15 (a) how the higher frequency oscillations are superimposed on the lower frequency motion.

When the current is in the z direction, the tower oscillates about $y \approx 0.4\ m$ and $z \approx 2.0\ m$, but now in the y direction as shown in Fig. 2.16 (b). The frequencies of oscillation are the same as those in Fig. 2.15, but the high frequency oscillations in Fig. 2.16 (a) are more pronounced.

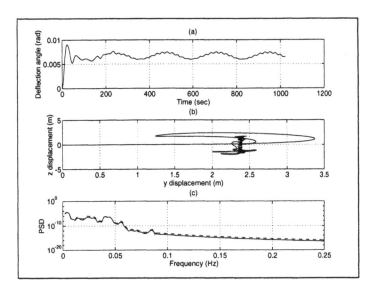

2.15 Equilibrium Position in the Presence of Current and Wind, $C_L = 1, \alpha = 0, \nu = 0$.
(a) θ Time Domain, (b) Tower's Top Motion, (c) θ Frequency Domain.

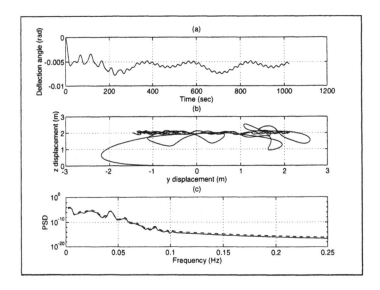

2.16 Equilibrium Position in the Presence of Current and Wind, $C_L = 1$,
$\alpha = 90^0, \nu = 0$. (a) θ Time Domain, (b) Tower's Top Motion, (c) θ Frequency Domain.

Wave, Current and Wind Excitation: In this part of the study, the wave height H is much smaller the mean water level d, that is, $H \ll d$. Therefore, the relation between the wave height and the wave frequency ω given in Hooft (1981) [61] is used,

$$\frac{Hk}{0.28\pi \tanh kd} = 1. \tag{2.165}$$

This relation, with the deep water simplification $\tanh kd = 1$, and $k = \omega^2/g$, leads to

$$\omega = \sqrt{\frac{g\pi}{3.5H}}. \tag{2.166}$$

The fluid parameter values used are $C_D = 1.2, C_M = 1.5$ and $C_L = 1.0$, with significant wave height $H = 3\,m$. First, the response due to waves without current and wind is depicted in Fig. 2.17. The deflection angle θ oscillates at the wave frequency given by equation (2.166), $\omega = 0.12\,Hz$ about the zero equilibrium position.

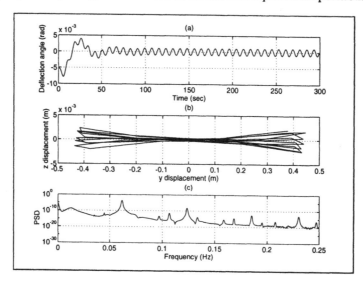

2.17 Response Due to Wave Excitation, $U_c = 0, u_w = 0, \alpha = 0, \nu = 0$ deg.

Fig. 2.18 shows the response due to combined waves, current and wind propagating in the y direction. The oscillations are now about an equilibrium position of $y \approx 2.4\,m$, and $z = 0$, which can be calculated from equation (2.161). The PSD contains several loading frequencies, as shown in Fig. 2.18 (c). These are the natural frequency $\omega_n = 0.028\,Hz$ and its multiples, the wave frequency $\omega = 0.12\,Hz$, the vortex shedding frequency due to waves $\omega_s = 2\omega = 0.24\,Hz$, and the vortex shedding frequency due to wind $\omega_w = 0.04\,Hz$.

The additional effect of wind in the z direction is depicted in Fig. 2.19. The tower oscillates at the same frequencies as for the case in Fig. 2.18, but now about an equilibrium position of $y \approx 2.2\,m$ and $z \approx 0.04\,m$.

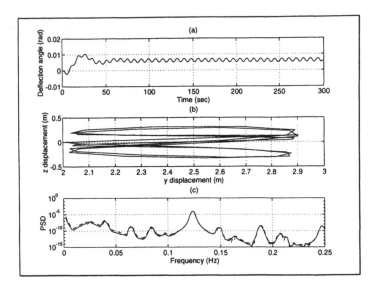

2.18 Response Due to Wave Excitation, $U_c = 1m/s$, $u_w = 20m/s$, $\alpha = 0, \nu = 0$ deg.

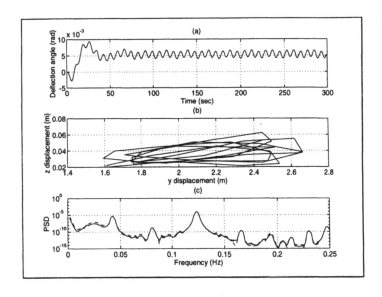

2.19 Response Due to Wave Excitation, $U_c = 1m/s$, $u_w = 20m/s$, $\alpha = 0, \nu = 90$ deg.

Finally, setting the vortex shedding coefficient to zero results in a response depicted in Fig. 2.20. The vortex shedding frequencies disappeared as expected and only two main frequencies remain, the natural frequency and the wave frequency. Thus, it is concluded that a variety of phenomena can be modeled, turned on or turned off. This is particularly valuable for ascertaining which loading is crucial and which not.

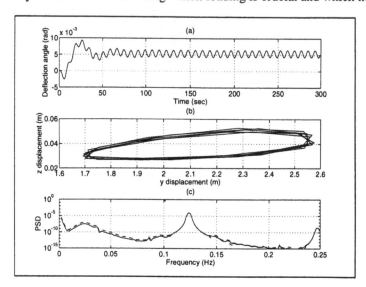

2.20 Response due to wave excitation with $C_L = 0$. $U_c = 1m/s$, $u_w = 20m/s$, $\alpha = 0, \nu = 0$ deg.

Resonance Response: Next the tower's response to harmonic, subharmonic and superharmonic excitation is investigated and compared to the response of the single degree of freedom model. Figs. 2.21 and 2.22 present the tower's response to wave excitation at the natural frequency of the system $\omega \approx \omega_n$. The wave height is $H = 1$ m and the drag coefficient $C_D = 0$. The time domain response, Fig. 2.21 (a), clearly shows the beating phenomenon. The frequency response is especially rich in content.

Fig. 2.22 depicts the tower response in the phase plane (a), and shows the tower's top motion (b). Had more data points been generated in the simulation, one would not see the "corners" in both (a) and (b), but rather smooth curves.

Subharmonic response is depicted in Figs. 2.23 and 2.24. Here the excitation frequency is nearly twice the natural frequency of the system $\omega \approx 2\omega_n$. In the time domain, it can be seen that the beating phenomenon is not as pronounced as in the harmonic response and the amplitude is much smaller. This can be attributed to there being less energy transfer where the loading frequency is twice the natural frequency. Again, the jaggedness of the curves are a function of the number of data points used in the simulation.

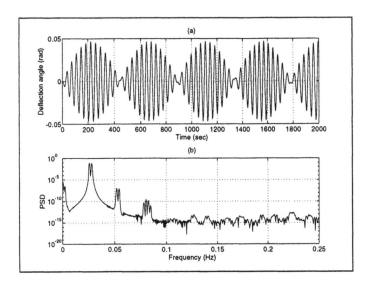

2.21 Undamped Response to Harmonic Excitation $\omega \approx \omega_n$. (a) Time Domain, (b) Frequency Domain.

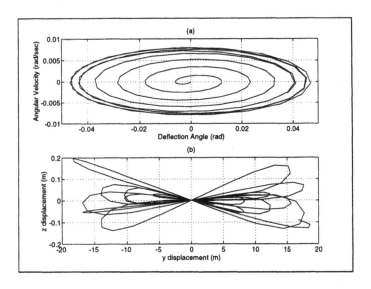

2.22 Undamped Response to Harmonic Excitation $\omega \approx \omega_n$. (a) Phase Plane, (b) Tower's Top Motion.

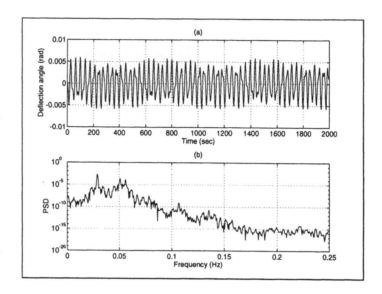

2.23 Undamped Response to Harmonic Excitation $\omega \approx 2\omega_n$. (a) Time Domain, (b) Frequency Domain.

From the phase plane (Fig. 2.24(a)), it is seen that the trajectories do not repeat, but investigating this point further showed it to be a quasiperiodic response and not a chaotic one.

Superharmonic response, in which the excitation frequency is at about one half of the natural frequency, $\omega \approx \frac{1}{2}\omega_n$, is depicted in Figs. 2.25 and 2.26. The amplitude of the response is small and there is no beating. The response, like in the subharmonics, is a quasiperiodic motion. Again, less energy is added at the half–natural frequency.

Finally, the harmonic response in the presence of drag force, with $C_D = 1.2$, is found. The beating phenomenon vanishes and the trajectories in the phase plane repeat as shown in Figs. 2.27 and 2.28. Again the damping dissipates the energy necessary for beating.

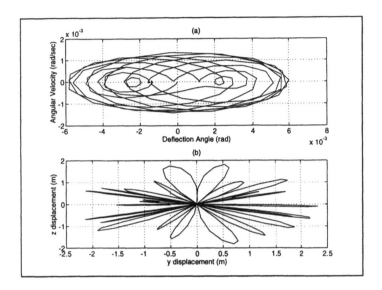

2.24 Undamped Response to Harmonic Excitation $\omega \approx 2\omega_n$. (a) Phase Plane, (b) Tower's Top Motion.

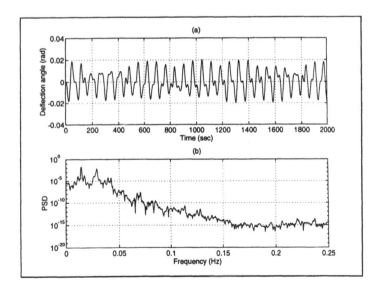

2.25 Undamped Response to Harmonic Excitation $\omega \approx \frac{1}{2}\omega_n$. (a) Time Domain, (b) Frequency Domain.

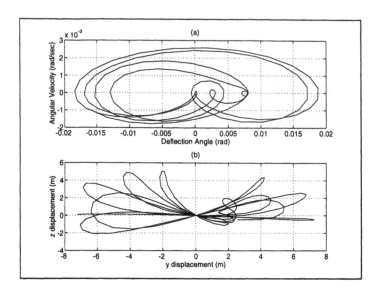

2.26 Undamped Response to Harmonic Excitation $\omega \approx \frac{1}{2}\omega_n$. (a) Phase Plane, (b) Tower's Top Motion.

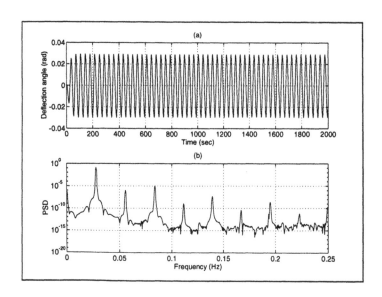

2.27 Response to Harmonic Excitation $\omega \approx \omega_n$ with Damping $C_D = 1.2$. (a) Time Domain, (b) Frequency Domain.

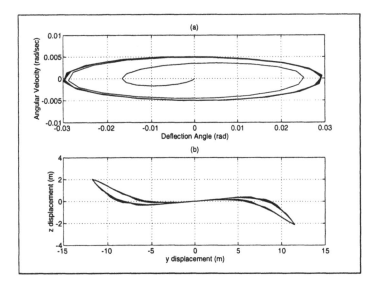

2.28 Response to Harmonic Excitation $\omega \approx \omega_n$ with Damping $C_D = 1.2$. (a) Phase Plane, (b) Tower's Top Motion.

Comparing the resonance responses for the two degree of freedom model with those of the single degree of freedom model (see Bar-Avi [34]), it is found that the regions about the resonance frequencies in which beating occurs, $(\frac{1}{2}\omega_n, \omega_n, 2\omega_n)$, are much smaller and less pronounced for the two degree of freedom model. This is because some of the energy is also transferred to the second coordinate φ, leaving less energy for the θ coordinate oscillation.

Response to Wave Slamming, Coriolis Acceleration and Wind: The tower's response due to wave slamming, Coriolis acceleration and wind loads are investigated next. Wave slamming and Coriolis acceleration, although having small influences on the response, are nevertheless important. Wave slamming, due to its high frequency content, results in very small response amplitudes of high frequency, and are therefore of concern to fatigue life studies. The Coriolis load has the effect of coupling the two degrees of freedom.

Typical wave slamming response is shown in Fig. 2.29. The significant wave height is $H = 5\ m$, the wave frequency is nearly the natural frequency of the tower, $\omega \approx \omega_n$, and all fluid parameters are set to zero. The very small amplitude motion is in the y direction only since there is no transverse force. As can be seen from the figure, the response beats since the frequency of the impulsive force is very close to the tower's natural frequency.

When Earth rotation is added to the model, a transverse (gyroscopic) moment causes a coupling between the two degrees of freedom so the response due to wave slamming is not planar any longer, as shown in Fig. 2.30(c).

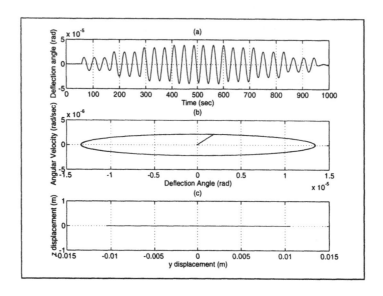

2.29 Response to Wave Slamming, $H = 5m$, $C_L = C_M = C_D = 0$. (a) Time Domain, (b) Phase Plane, (c) Tower's Top Motion.

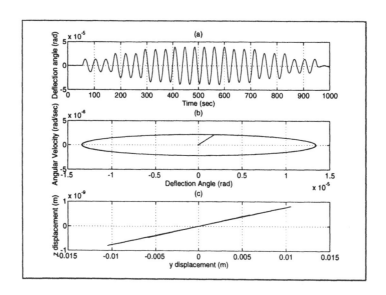

2.30 Response to Wave Slamming and Earth Rotation, $H = 5m$, $C_L = C_M = C_D = 0$. (a) Time Domain, (b) Phase Plane, (c) Tower's Top Motion.

Fig. 2.31 shows the response due to waves and Coriolis effect. The vortex shedding coefficient $C_L = 0$, but as can be seen from the figure, there is a transverse force (like the vortex shedding) caused by the rotation of the Earth.

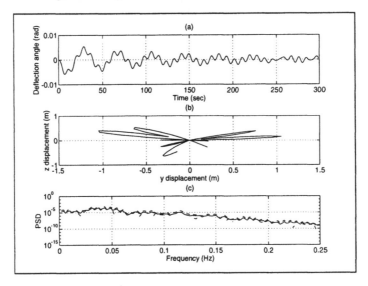

2.31 Response due to Waves and Coriolis Acceleration.

Wind has a larger effect on the response than do slamming and Coriolis loads. Figs. 2.32 and 2.33 depict the tower's response due to wind in the y and z directions, respectively. One can therefore conclude as to the importance of the various loads, but as it is also known, small loads can have large effects, such as fatigue.

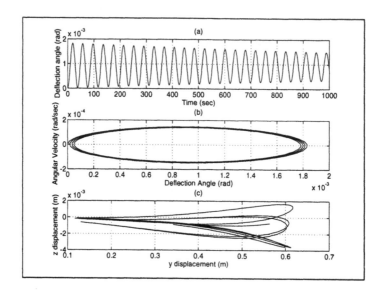

2.32 Influence of Wind Load, $\nu = 0$. (a) Time Domain, (b) Phase Plane, (c) Tower's Top Motion.

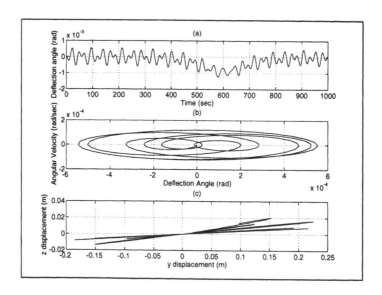

2.33 Influence of Wind Load, $\nu = 90^0$. (a) Time Domain, (b) Phase Plane, (c) Tower's Top Motion.

Chaotic Response: Due to the nonlinearity in the model, chaotic response is expected to be possible, although not highly probable. Investigation of the chaotic response, and mapping its regions of occurrence, are beyond the scope of this study. Nevertheless for completeness, a chaotic response is shown. Fig. 2.34 depicts the tower response for wave height $H = 5\ m$ with $\omega = 0.1\ rad/s$ and zero initial conditions. The trajectories in the phase plane do not repeat.

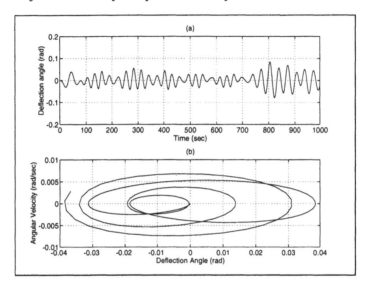

2.34 Chaotic Response, with Zero Initial Conditions, $\theta_0 = 0, \dot{\theta}_0 = 0$. (a) Time Domain, (b) Phase Plane .

Fig. 2.35 depicts the same response but with *nonzero* initial conditions. The responses are different, as can be seen by comparing the figures.

The proof that the response is chaotic is given by the Poincare' map that is shown in Fig. 2.36. For $C_D = 0$, the points in Fig. 2.36(a) are scattered in an erratic fashion, but for $C_D = 1.2$, the points in Fig. 2.36(b) are much more organized, almost as in a quasiperiodic response.

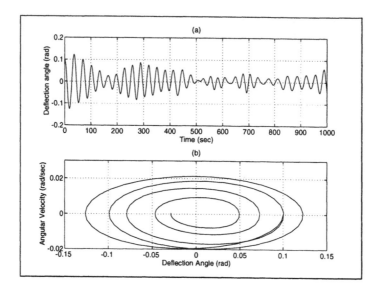

2.35 Chaotic Response, with Non–Zero Initial Conditions, $\theta_0 = 0.1rad$, $\dot{\theta}_0 = 0$. (a) Time Domain, (b) Phase Plane.

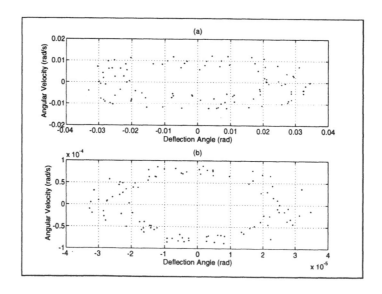

2.36 Poincare' Map of the Chaotic Response. (a) $C_D = 0$, (b) $C_D = 1.2$.

Stochastic Response

The stochastic response of the tower due to uniformly distributed random parameters is evaluated utilizing Monte–Carlo simulations. The governing nonlinear differential equations of motion (2.159) and (2.160) are repeatedly solved using 'ACSL'. At each run, different values are assigned to the parameters via the probability distribution function, and the average response is calculated until a convergence is achieved, that is, the change in the averages between the current and previous runs is less than 1%. The assumed probability distributions for the fluid coefficients, the wave height, and the wave frequency used in the simulations are taken from Hogben et al. 1977 [62] and are shown in Table 2.3.

μ	C_D	C_M	C_L	H	α^0
0.05 to 0.2	0.6 to 2.0	0.6 to 2.0	1.4 to 2.0	1 to 3	$0 \pm 10, 90 \pm 10$

2.3 Uniformly Distributed Random Parameters.

The tower's average deflection angles θ_{av} and φ_{av}, and their bounds, $\theta_{av} \pm \sigma$ and $\varphi_{av} \pm \sigma$, where σ is the respective standard deviation, are calculated and plotted for different parameters. Also, the stochastic response due to waves with random height, governed by the Pierson–Moskowitz spectrum, is evaluated.

Random Fluid Parameters: In this run all parameters are kept constant except for the fluid constants C_D, C_M, and C_L which were assumed random. First we consider the free vibration. The wave height is set to zero, and an initial velocity on the deflection angle is taken to be $\dot{\theta}(t = 0) = 0.001$ *rad/s*.

Fig. 2.37 shows the deflection angle (a) in the time domain and (b) in the frequency domain for the mean response $\theta_{av}(t)$ and the bounds $\theta_{av}(t) \pm \sigma(t)$. The parameter that causes a change in the natural frequency is C_M because it is related to the tower's moment of inertia through $C_A = C_M - 1$. The average fundamental frequency is about $\omega_n = 0.028$ *Hz*.

Next, waves are included. The wave height is taken as $H = 2$ *m*, the wave frequency $\omega = 0.5$ *rad/s*, the Coulomb friction coefficient $\mu = 0$, and the current velocity $U_c = 0$. Fig. 2.38(a) depicts the average deflection θ_{av} of the tower with bounds of $\theta_{av} \pm \sigma$, and Fig. 2.38(b) depicts the average rotation angle φ_{av} and $\varphi_{av} \pm \sigma$. From the figure we see that the average steady state deflection θ_{av} is about zero since the current velocity is zero, and the average is bounded, -0.005 *rad* $< \theta_{av} < 0.005$ *rad*. The rotation average angle φ_{av} is much larger and grows continuously with time, as expected.

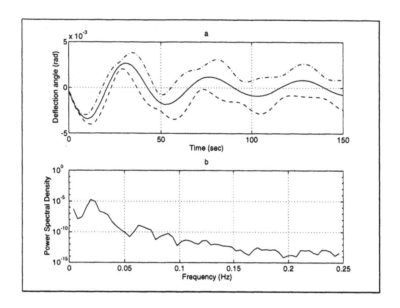

2.37 Damped Free-Vibration Response. (a) Time Domain, (b) Frequency Domain.

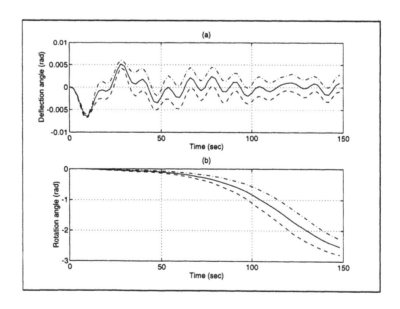

2.38 Tower Response to Random Fluid Parameters. (a) Deflection Angle θ_{av} (solid line), $\theta_{av} \pm \sigma$ (dashed line), (b) Rotation Angle φ_{av} (solid line), $\varphi_{av} \pm \sigma$ (dashed line).

The tower's end displacement in the z and y directions is plotted in Fig. 2.39(a). It can be seen that the displacement in the y direction is larger than that in the z direction since the wave propagates in this direction. Also, the motion is oscillatory about the zero position. Fig. 2.39(b) shows the frequency content of the displacements (y, z).

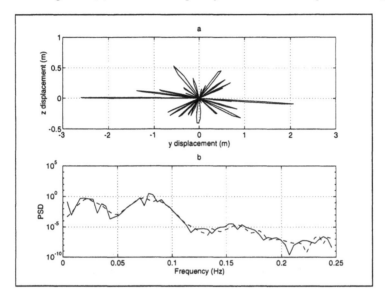

2.39 Tower Response to Random Fluid Parameters. (a) Tower's End Displacements, (b) Tower's Response in the Frequency Domain; y–solid line, z–dashed line.

Influence of Current Direction: The influence of current direction is next investigated. The fluid parameters are constant, $C_D = 1.2$, $C_M = 1.5$ and $C_L = 1.0$. All other parameters are the same as in the previous run. Fig. 2.40 shows the tower's response for current direction of $\alpha = 0^0 \pm 10^0$. This shorthand notation signifies that the mean direction of $\alpha = 0^0$ is used along with a uniform distribution that ranges between $\alpha = -10^0$ and $\alpha = +10^0$. It can be seen that both oscillations fluctuate about an equilibrium position which is not zero. The standard deviation of the average rotation angle φ_{av} is larger than the average deflection angle θ_{av}, primarily due to the fact that the rotation angle φ_{av} can be much larger than the deflection angle.

Fig 2.41(a) depicts the tower's end displacement and (b) its frequency responses. The tower's steady state motion (after the transient decays) oscillates about an equilibrium position of $(y, z) = (5 \pm 0.5, 4.6 \pm 0.1)\ m$. The shift of the equilibrium position is due to current. From the frequency responses in (b), the wave frequency $\omega = 0.08$ Hz and the vortex–shedding frequency $2\omega = 0.16\ Hz$ are clearly seen.

The responses for $\alpha = 90^0 \pm 10^0$ are shown in Figs. 2.42 and 2.43. The tower's steady state response oscillates about an equilibrium position of $(y, z) = (-4.5 \pm 0.5, 5.5 \pm 0.2)\ m$. The change in the equilibrium position is due to current direction.

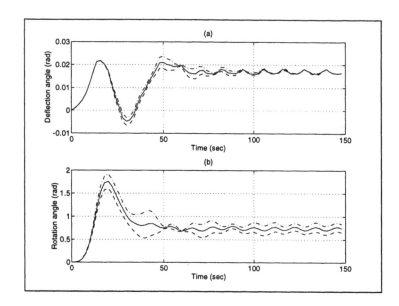

2.40 Influence of Current Direction - $\alpha = 0^0 \pm 10^0$. (a) Deflection Angle θ_{av} (solid line), $\theta_{av} \pm \sigma$ (dashed line), (b) Rotation Angle φ_{av} (solid line), $\varphi_{av} \pm \sigma$ (dashed line).

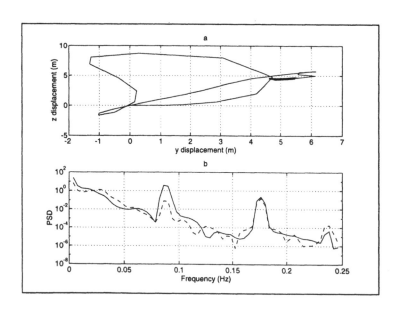

2.41 Influence of Current Direction - $\alpha = 0^0 \pm 10^0$. (a) Tower's End Displacements, (b) Tower's Response in the Frequency Domain; y-solid line, z-dashed line.

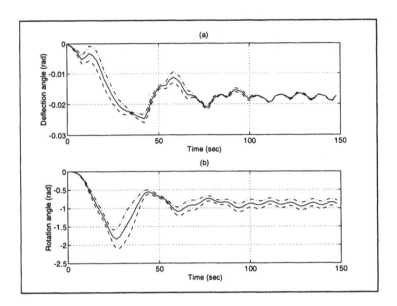

2.42 Influence of Current Direction - $\alpha = 90^0 \pm 10^0$. (a) Deflection Angle θ_{av} (solid line), $\theta_{av} \pm \sigma$ (dashed line), (b) Rotation Angle φ_{av} (solid line), $\varphi_{av} \pm \sigma$ (dashed line).

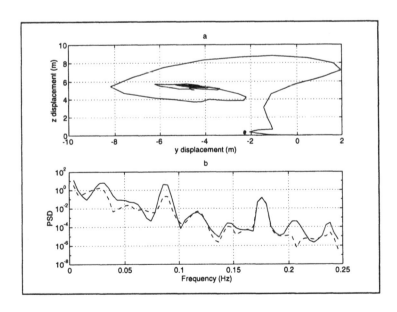

2.43 Influence of Current Direction - $\alpha = 90^0 \pm 10^0$. (a) Tower's End Displacements, (b) Tower's Response in the Frequency Domain; y-solid line, z-dashed line.

Reducing the drag coefficient to $C_D = 0.6$ changes the tower's response as can be observed from Fig. 2.44. The equilibrium position is different than that with $C_D = 1.2$ due to the fact that the position depends on $C_D |U_c \cos \alpha| \, U_c \cos \alpha$, as explained in the single degree of freedom solution in Bar-Avi and Benaroya [33] . Large standard deviations can be observed for φ.

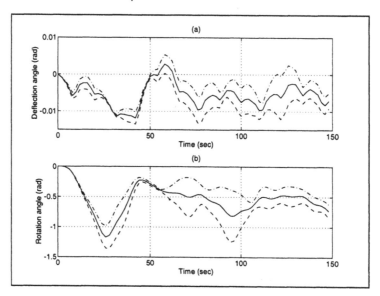

2.44 Influence of Current Direction - $\alpha = 0^0 \pm 10^0$. (a) Deflection Angle θ_{av} (solid line), $\theta_{av} \pm \sigma$ (dashed line), (b) Rotation Angle φ_{av} (solid line), $\varphi_{av} \pm \sigma$ (dashed line).

The influences of the current direction and the drag coefficient on the average response is summarized in the next two figures. Fig. 2.45(a) depicts the average deflections θ_{av} and Fig. 2.45(b) shows the average rotation angles φ_{av} for mean current direction $\alpha = 0^0$ and 90^0 with constant drag coefficient $C_D = 1.2$. It can be seen that for $\alpha = 0^0$, the average response is $\theta_{av} = 0.02 \; rad$, and $\varphi_{av} = 0.7 \; rad$, while for $\alpha = 90^0$, $\theta_{av} = -0.02 \; rad$ and $\varphi_{av} = -0.85 \; rad$. The direction of the current causes a change in the direction of the lift force, which results in different equilibrium positions.

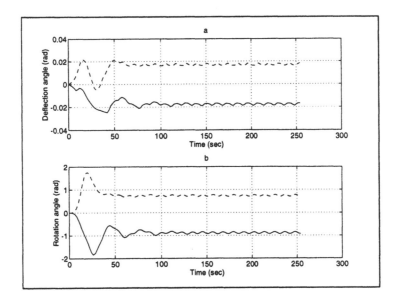

2.45 Tower's Average Response for $\alpha = 0^0 \pm 10^0$ (dashed line), and $\alpha = 90^0 \pm 10^0$ (solid line) with $C_D = 1.2$. (a) deflection angle θ_{av}, (b) rotation angle φ_{av}.

Friction, Waves Slamming and Earth Rotation: The influence of uncertainty in Coulomb friction on the average steady state response θ_{av} and φ_{av} is very small. As can be seen in Fig. 2.46, the transient response is smaller with friction, but the steady state is essentially the same.

The tower response to random wave slamming in the presence of Earth rotation is shown in Figs. 2.47 and 2.48. It can be seen that the effects are very small, as was shown for the deterministic response. The standard deviation for both angles are negligible due to the very low response. Also, although the system is nonlinear, only the natural frequency can be seen in the Fig. 2.48(b), again due to the very low response. The figures show how the slamming force is "turned on" at about 160 *s*.

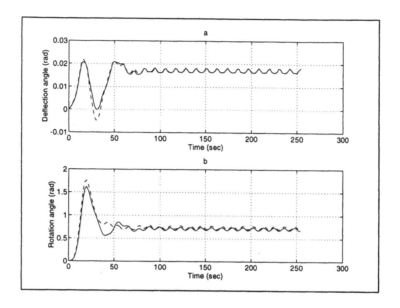

2.46 Influence of Coulomb Friction on the Tower's Average Response. $\mu = 0$ (deterministic– solid line), $\mu = 0.05$ to 0.2 (random dashed line).

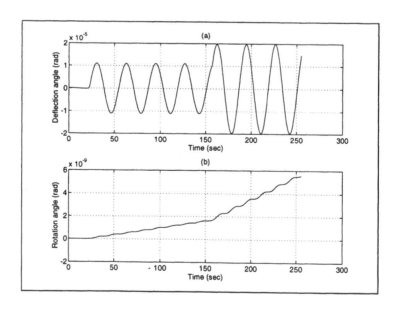

2.47 Influence of Wave Slamming and Earth Rotation. (a) Deflection Angle θ_{av} (solid line), $\theta_{av} \pm \sigma$ (dashed line), (b) Rotation Angle φ_{av} (solid line), $\varphi_{av} \pm \sigma$ (dashed line).

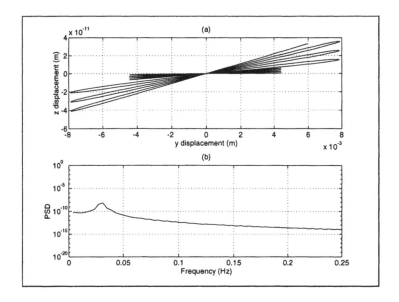

2.48 Influence of Wave Slamming and Earth Rotation. (a) Tower's End Displacements,
(b) Tower's Response in the Frequency Domain; y-solid line, z-dashed line.

Influence of Wave Height: Figs. 2.49 and 2.50 describe the tower's response with
parameters H, C_D, C_M, C_L, and μ modeled as random uniformly distributed variables
with zero current velocity. Fig. 2.49 shows θ_{av} and $\theta_{av} \pm \sigma$, φ_{av} and $\varphi_{av} \pm \sigma$, while Fig.
2.50(a) depicts the average displacement of the tower's end and Fig. 2.50(b) shows the
average response in the frequency domain. It may be argued that gravity wave loading
is the single most important environmental force on the offshore structure. This does
not mean that there are no larger forces on occasion, but rather that the gravity waves
are always there, can vary widely in their amplitudes, and have a cumulative effect on
the structure.

Next, the wave height is assumed random and governed by the Pierson–Moskowitz
spectrum. The tower's response to two different significant wave heights is evaluated.
Figs. 2.51 and 2.52 depict the responses for a significant wave height of $H_s = 4\ m$.

Figs. 2.53 and 2.54 show the same responses for $H_s = 15\ m$. By comparing the
two responses, it can be seen from the phase plane that although the responses are of the
same order of magnitude, the motions are different. This is due to different frequency
content for different significant wave height.

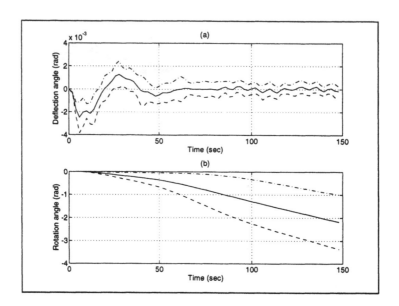

2.49 Tower Response to Random Parameters. (a) Deflection Angle θ_{av} (solid line) $\theta_{av} \pm \sigma$ (dashed line), (b) Rotation Angle φ_{av} (solid line), $\varphi_{av} \pm \sigma$ (dashed line).

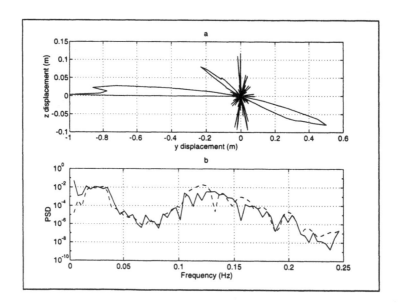

2.50 Tower Response to Random Parameters. (a) Tower's End Displacements, (b) Tower's Response in the Frequency Domain; y-solid line, z-dashed line.

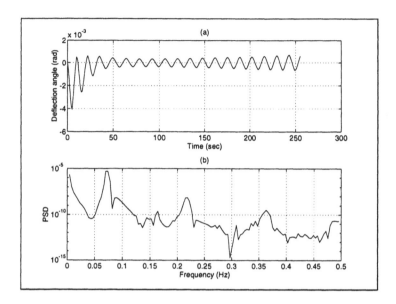

2.51 Influence of Random Wave Height, $H_s = 4m$. (a) Deflection Angle in Time Domain, (b) Deflection Angle in Frequency Domain.

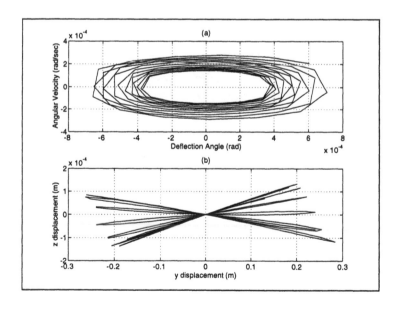

2.52 Influence of Random Wave Height, $H_s = 4m$. (a) Deflection Angle in Phase Plane, (b) Tower's Top Motion.

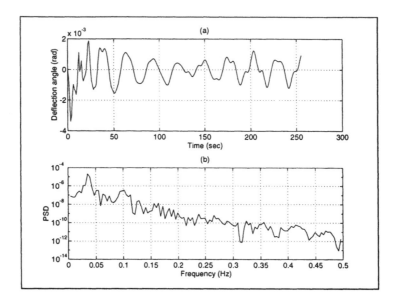

2.53 Influence of Random Wave Height, $H_s = 15m$. (a) Deflection Angle in Time Domain, (b) Deflection Angle in Frequency Domain.

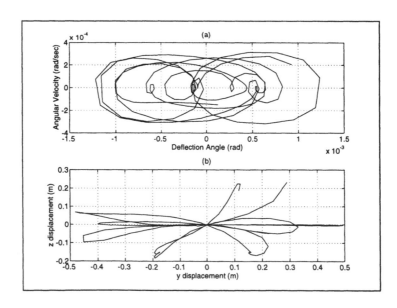

2.54 Influence of Random Wave Height, $H_s = 15m$. (a) Deflection Angle in Phase Plane, (b) Tower's Top Motion.

Response of an Appendaged Tower

In this section the response of an elastic appendaged tower, such as an antenna or a mast connected to the deck of an articulated tower, is investigated. It is assumed that the inertia of the mast is very small compared to that of the articulated tower to which it is connected. Hence, the mast has no effect on the tower. The mast is fixed at the bottom and has a concentrated mass at its top. It is subjected to deterministic and random wind loads as well as horizontal and vertical base excitation. These motions are derived from the response of the articulated tower that was investigated in Sections 3.3 and 3.4. The following physical parameters are used in the analysis,

- Mast's length $l = 35\ m$
- Mast's diameter is tapered, $D = 0.3\ m$ at the bottom, to 0.2 m at the top
- Young's Modulus $E = 2.04 \times 10^{11}\ N/m^2$
- Mast density $\rho_T = 7800\ Kg/m^3$
- End mass $M = 250\ Kg$
- Deterministic wind mean speed at 10 m, $U_{10} = 10$ to 20 m/s
- Gust wind mean speed at 10 m, $U_{Gust} = 15\ m/s$
- Wind drag coefficient $C_{D_a} = 1.4$.

The analysis includes free vibration, response due to base excitation, response due to deterministic and random wind speed, and the combination of both. Many types of random wind speed spectra exist in the literature. Among them are those due to Davenport, Harris, Kareem and others. See Chakrabarti [60] pp. 101. In the investigation presented here, the Davenport spectrum is assumed (Patel [9] pp. 187),

$$S_{vv}(f) = \frac{4k\bar{f}}{(1 + \bar{f}^2)^{4/3}} \frac{U_{Gust}^2}{f}, \qquad (2.167)$$

where f is the loading frequency in Hz, U_{Gust} is the mean gust wind velocity in m/s, k is the sea surface drag coefficient (taken to be 0.005), and $\bar{f}$ is a normalized frequency given by

$$\bar{f} = \frac{fL}{U_{Gust}}, \qquad (2.168)$$

with L being a representative length scale, taken to be 1200 m.

To numerically solve the equation of motion, the wind speed spectrum must be transformed from the frequency domain into time domain. Borgman's method was used to transform the Pierson-Moskowitz spectrum, but could not be utilized here due to a major difference between the Pierson-Moskowitz spectrum and the Davenport spectrum. Hence a brute force approximation is used. The spectrum is approximated as a series of sine functions with amplitudes that are calculated using equation (2.167),

$$U_w(t) = \sum_{i=1}^{n} A_i \sin(f_i t), \qquad (2.169)$$

where

$$A_i = 2U_{Gust}\sqrt{\frac{k\bar{f}_i}{(1+\bar{f}_i^2)^{4/3}}},\qquad(2.170)$$

and n being the number of divisions of the frequency span. Fig. 2.55 shows the actual Davenport spectrum and the approximated one. As can be seen, they agree quite well.

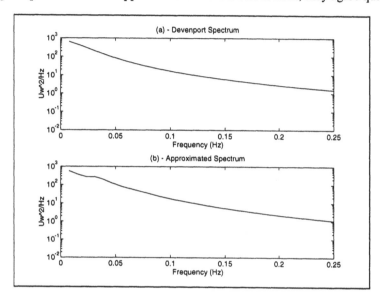

2.55 A Comparison Between the Actual Davenport Spectrum and the Approximated Spectrum.

First, the natural frequencies of the mast are found from the free vibration response, depicted in Fig. 2.56. The figure shows the response of the top end of the mast in time and frequency domains with structural damping $\zeta = 0.04$ (Figs. c,d), and without damping (Figs. a,b). The first peak corresponds to the first fundamental frequency, $\omega_1 = 0.13\ Hz$, and the second fundamental frequency is $\omega_2 = 1.12\ Hz$. The peaks in between are multiples of the first natural frequency, due to the nonlinearity of the system. From a linear eigenvalue analysis performed using 'ACSL', the same two fundamental frequencies were found.

Constant wind causes the mast to change its equilibrium position, just as the current does to the articulated tower. Since the wind force is proportional to the square of the wind speed, the mast's equilibrium position is also proportional to the square of the wind speed. Fig. 2.57 depicts the deflection of the mast's top end, when exposed to wind speed of $U_{10} = 10\ m/s$ and $U_{10} = 20\ m/s$. It is clearly seen that the response due to wind speed of 20 m/s (solid line) is four times larger than the one for wind speed of 10 m/s (dashed line). The response of the whole mast to a wind speed of 20 m/s is depicted in Fig. 2.58.

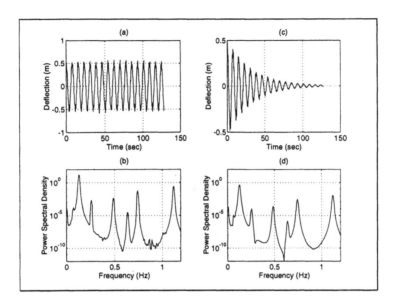

2.56 Free Vibration Response in Time and Frequency Domains. (a,b) Without Damping, (c,d) With Structural Damping, $\zeta = 0.04$.

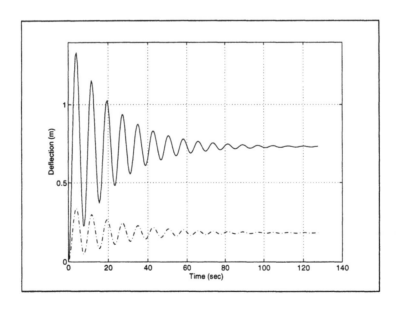

2.57 Equilibrium Position in the Presence of Wind. $U_{10} = 20 \; m/s$ - Solid Line, $U_{10} = 10 \; m/s$ - Dashed Line.

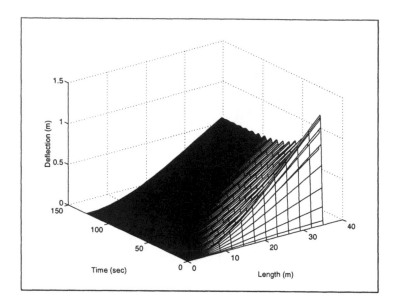

2.58 Equilibrium Position of the Whole Mast Due to Wind Speed of $U_{10} = 20\ m/s$.

The response of the mast's top to constant wind speed, random wind speed and a combination of both, is depicted in Fig. 2.59. The solid line is the mast's response to a deterministic wind speed of 20 m/s. The dashed line is for a random wind speed having a mean speed of 15 m/s, and the dashed solid line is the response due to both together. The first natural frequency of the mast is clearly seen from the response in the frequency domain. The response of the whole mast due to both random and deterministic wind speed is described in Fig. 2.60

The response due to base excitation is investigated next. As mentioned earlier, the base excitation is derived using the solution of a partially submerged articulated tower (Section 3.4). From the deflection angle $\theta(t)$ of the rigid tower due to waves, the horizontal u and vertical w base motions are found,

$$w = 400 \sin \theta(t) \tag{2.171}$$
$$u = 400(1 - \cos \theta(t)). \tag{2.172}$$

The horizontal motion u and its time derivatives have the same effect on the appendage as does an external force, while the vertical base excitation w and its derivative become part of the equation of motion and, for certain amplitudes and frequencies, can cause parametric instabilities, as will be shown later.

The response of the mast to deterministic wave excitation having a height of $H = 3$ m and frequency $\omega = 0.12\ Hz$ is shown in Fig. 2.61. The time domain response of the top end of the mast is seen to beat because the excitation frequency is very close to the first natural frequency of the system $\omega_1 = 0.128\ Hz$. Fig. 2.62 depicts the response of the whole mast.

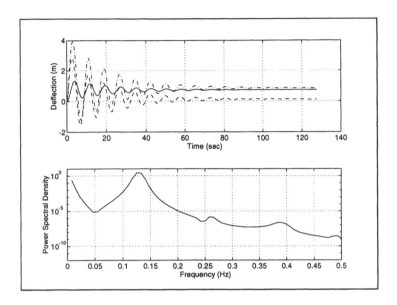

2.59 Response to Wind Force Excitation in Time and Frequency Domains. Solid Line - $U_{10} = 20$ m/s, Dashed Line - $U_{Gust} = 15$ m/s, Dashed-Dotted Line - Both.

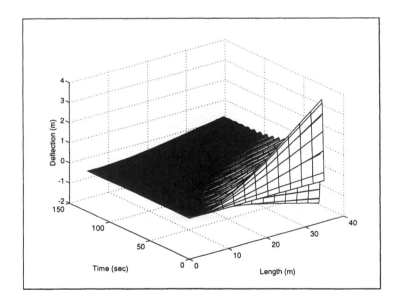

2.60 Mast's Response to Deterministic plus Random Wind Excitation.

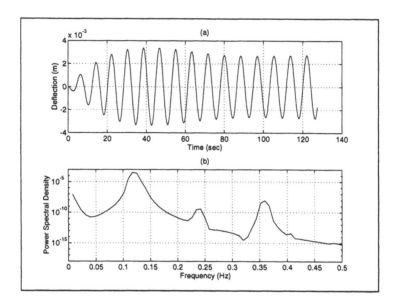

2.61 Time and Frequency Domain Responses of the Top End of the Mast Due to Deterministic Wave Height.

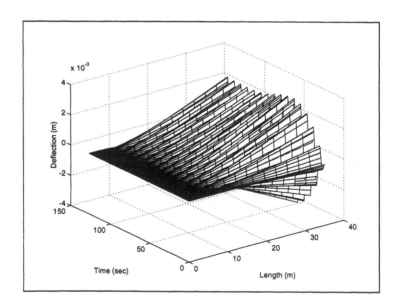

2.62 Response of the Whole Mast to Deterministic Wave Excitation.

When the frequency of the base excitation coincides with the first natural frequency of the structure, instability occurs, as can be seen from Fig. 2.63. In the absence of damping, a very low amplitude load can cause instability, but when damping is added to the system, a larger amplitude load is needed to cause instability.

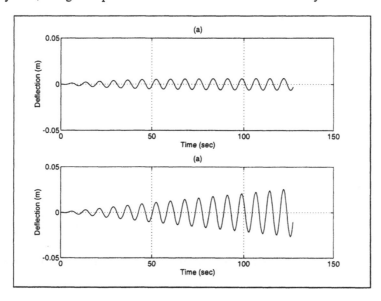

2.63 Parametric Instability Due to Vertical Base Excitation at the Natural Frequency. (a) With Structural Damping $\zeta = 0.04$, (b) No Damping.

The motion of the top end of the mast to random wave excitation having a significant height of $H_s = 4\ m$ is shown in Fig. 2.64, with Fig. 2.65 showing the response of the whole mast. Because the excitation frequency of the deterministic wave is very close to first natural frequency of the mast, the response to deterministic waves is larger than the random response.

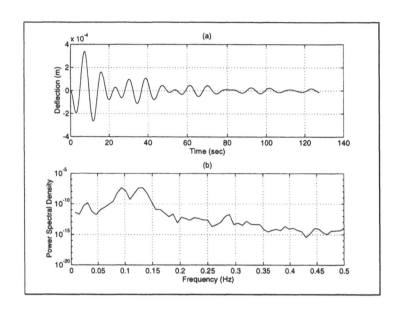

2.64 Time and Frequency Domain Response of the Top End of the Mast Due to Random Wave Height.

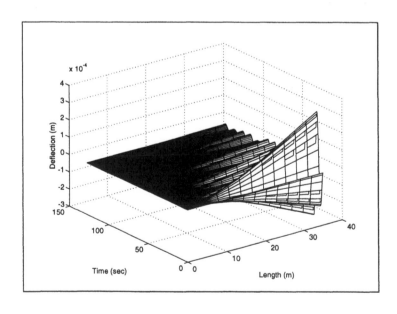

2.65 Response of the Whole Mast to Random Wave Excitation.

The dynamic response of the mast to wind and wave excitation is shown next. Figs. 2.66 and 2.67 depict the response of the mast due to deterministic wave and wind excitation. The wind speed is set to $U_{10} = 20$ *m/s* and the waves are of $H = 3$ *m* and $\omega = 0.12$ *Hz*. Fig. 2.66 shows the time and frequency domain responses of the top end, and Fig. 2.67 shows the whole mast response to deterministic wave and wind.

The response of the mast to random waves and wind is depicted in the following two figures. Fig. 2.68 shows the top end motion and Fig. 2.69 shows the response of the whole mast. The wind speed is $U_{Gust} = 15$ *m/s* and the significant wave height is set to $H_s = 4$ *m*.

By comparing the results of the response due to wind and wave, to those due to only wind, it can be seen that the base excitation response is negligible compared to that due to the wind. Base excitation plays an important role only when parametric instability occurs.

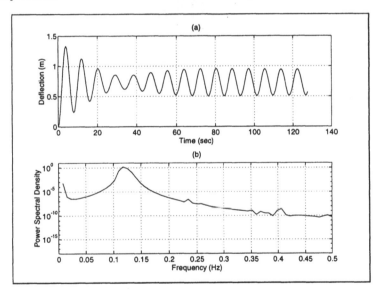

2.66 Time and Frequency Domain Response of the Top End of the Mast Due to Deterministic Waves and Wind Excitation.

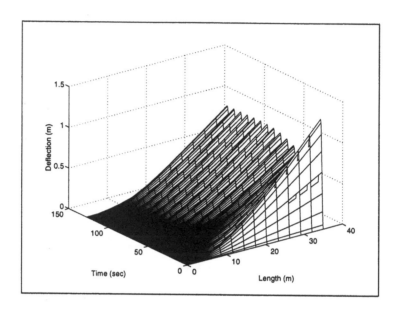

2.67 Response of the Whole Mast to Deterministic Waves and Wind Excitation.

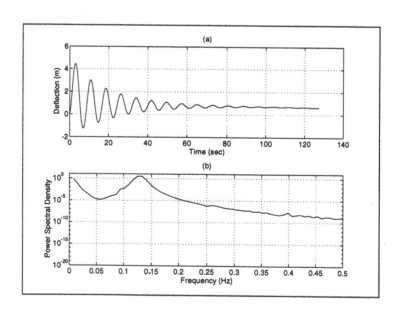

2.68 Time and Frequency Domain Responses of the Top End of the Mast Due to Random Waves and Wind Excitation.

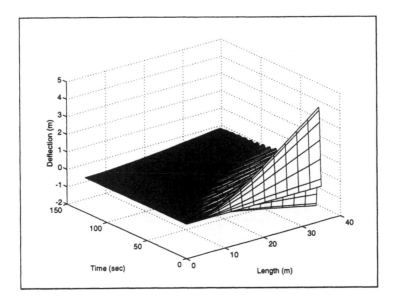

2.69 Response of the Whole Mast to Random Waves and Wind Excitation.

The equivalent stress along the mast is calculated using the Von-Mises formula, equation (1.134). The equivalent stress due to deterministic waves and wind excitation (same parameters as in Fig. 2.66) is depicted in Fig. 2.70, and the stresses due to random waves and wind (same parameters as in Fig. 2.68) is shown in Fig. 2.71.

Although this section is primarily an exercise, it demonstrates how the models developed herein can be utilized in a risk and reliability framework. Cycles can be counted, exceedances of certain stress levels can also be added for a measure of damage accumulation.

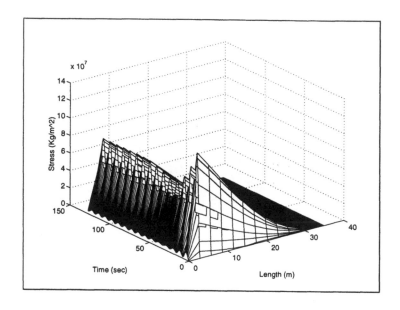

2.70 Equivalent Stress Due to Deterministic Waves and Wind.

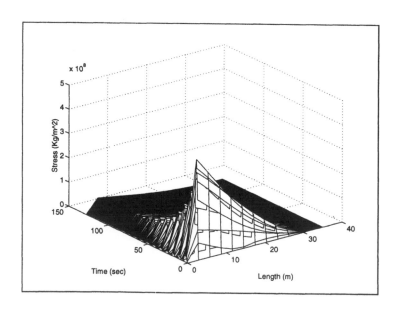

2.71 Equivalent Stress Due to Random Waves and Wind.

Summary

The nonlinear differential equations of motion for a two degree of freedom articulated tower partially submerged in the ocean are derived using Lagrange's equations. The tower is assumed to have the same dynamic properties as an upright spherical pendulum with additional effects and forces;

- Coulomb friction in the pivot (hinge)
- Structural viscous damping
- Gyroscopic moments due to the rotation of the Earth
- Drag fluid force due to waves and current, coupled to the structure
- Inertia and added mass fluid forces
- Drag wind load on the exposed part of the tower
- Wave slamming force
- Vortex shedding loads due to waves and wind
- Buoyancy force.

All fluid forces mentioned above due to waves, current, and wind are determined at the instantaneous position of the tower, resulting in two coupled and highly nonlinear ordinary differential equations with time–dependent coefficients.

The investigation of the two degree of freedom model presented in this chapter is divided into deterministic and stochastic analyses. In the deterministic analysis, the influences on the response of all forces mentioned above have been investigated and the following observations were drawn;

- The equilibrium position depends on the current and wind magnitude and direction. An analytical expression that matches the numerical results is found.
- The resonance response for harmonic, subharmonic and superharmonic loading is evaluated and the beating phenomenon is shown. The regions in which the beating occurs are very small and are not as pronounced as in the single degree of freedom model.
- The system can exhibit chaotic behavior depending on the wave's frequency and amplitude.
- The response to wave slamming is very small since the impulsive force is attenuated when the pulse period is shorter than the system fundamental period, which is the case here. It also beats since the force's period is the same as the tower's natural period.
- The Coriolis acceleration moment has a small but important influence on the response since it causes a coupling between the two degrees of freedom so that planar motion is not possible under real conditions.

In the stochastic analysis, the average response and the standard deviation due to uniformly distributed random parameters is evaluated. The equations are solved numerically for uniformly distributed random parameters such as wave height, current direction, fluid coefficients and Coulomb friction coefficient. Monte–Carlo simulations are performed, using 'ACSL', to determine the tower's average response and standard deviation. MATLAB is used for the frequency domain analysis.

From the analysis it is found that the standard deviation of the rotation angle φ_{av} is larger than that of the deflection angle θ_{av}. The average equilibrium position $(\theta_{av}, \varphi_{av})$ depends on the drag coefficient, and current velocity and direction. Coulomb friction is found to have a very small effect on the average steady state response, but a larger one on the transient response. The tower oscillates (sway) about the equilibrium position.

The response of a mast connected to the deck of an offshore articulated tower is then investigated. In this analysis the mast is assumed to be fixed to the bottom and carrying a concentrated mass at the top. The mast is subjected to deterministic and random wind forces, as well as to base excitation. The horizontal and vertical base excitation motions are derived from the motion of the articulated tower obtained in Chapter 3.

First, the equilibrium position due to wind is found. Wind forces are similar to current forces and they cause the mast to deflect to a stationary equilibrium position. The deflection of the mast is proportional to the air drag coefficient and to the square of the wind speed.

To find the response due to random wind velocity, the Davenport spectrum is used. First, the spectrum is transformed into a time history function, which is essentially a sum of sine functions, with varying amplitudes and frequencies. The response to random wind velocity with gusting mean speed $U_{Gust} = 15$ m/s is larger than that for a deterministic wind speed with mean velocity $U_{10} = 20$ m/s.

The base excitation has a horizontal component that acts like an external force, and a vertical one that acts as a parametric excitation. Although the response of the mast to base excitation is very small compared to the response due to wind, it is very important because for certain frequency and amplitude parameters, it can cause instability.

Finally, the stresses along the mast are evaluated, and it is found that a combination of deterministic wind speed $U_{10} = 20$ m/s and random wind speed with a mean gust $U_{Gust} = 15$ m/s results in a maximum equivalent stresses $\sigma_{eq} = 1.5 \times 10^8$ N/m^2 at the base of the mast. The yield stress of steel bar in tension is $\sigma_y = 2.8 \times 10^8$ N/m^2. It is concluded that wind forces are very important in analyzing these kind of structures.

In the next chapter, the basic ideas used to develop the articulated tower model will be extended and specialized for application to a tension leg platform.

3
Tension Leg Platforms

Of the classes of offshore structures, the Tension Leg Platform (TLP) is particularly well suited for deepwater operation(see Adrezin Bar-Avi and Benaroya [63]). Unlike fixed structures, its cost does not dramatically increase with water depth. The TLP is vertically moored at each corner of the hull minimizing the heave, pitch and roll of the platform. The resulting small vertical motion allows for less expensive production equipment than would be required on a semi–submersible [64] ,[65] . This structure, as opposed to the guyed tower, cannot be assumed to be a rigid body, and continuous elastic models have to be considered.

Design - General

Components

TLPs are complete oil and natural gas production facilities costing $1 billion or more [66] . The supporting structure of a TLP consists of a hull, tendons and templates, as shown in Fig. 3.1. The hull is a buoyant structure with a deck at its top that supports the oil production facility and crew housing. Pontoons and columns provide sufficient buoyancy to maintain the deck above the waves during all sea states. These columns are moored to the seafloor through tendons, and fixed in place with templates. The hull's buoyancy creates tension in the tendons.

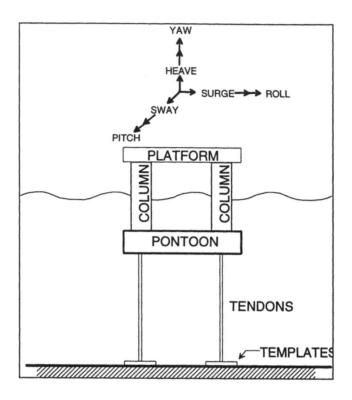

3.1 Schematic of a Tension Leg Platform (TLP).

The tallest TLP at the time of its construction, Shell Oil's Auger TLP in the Gulf of Mexico, began production in 1994 after an investment of six years and $1.2 *billion*. The Auger TLP with a crew of 112 has two main decks (300 by 300 ft^2) with a well bay at its center. Four cylindrical columns (74 ft diameter) and pontoons (28 by 35 ft^2 cross section) comprise the hull. There are three tendons at each column. Each tendon, also known as a tether or tension leg, was assembled from 12 steel pipes connected end to end, each with a 26 *in* diameter and a 1.3 *in* wall thickness, and a total length of 2,900 ft During severe storms the TLP may surge 235 ft [?] [?] .

Sato et al. (1989) [?] developed a basic planning method to design TLPs based on a customer's requirements. It begins with the design of the deck dimensions and layout and the determination of the weight of the on-board equipment. An initial hull displacement is then assumed and the hull particulars such as column displacement, pontoon diameter and air gap are designed. These designs are based on numerical analysis and tank testing. The actual displacement is then computed and compared with the initial assumed displacement. Adjustment to the hull displacement is made, and the process is repeated until all design criteria are satisfied.

The number of tendons and their dimensions are determined from the hull buoyancy, water depth and oceanic meteorological conditions. A high tendon outer diameter to wall thickness ratio is used to reduce the loss of payload due to tendon weight. The template design is dependent on the hull displacement and type of soil on the seafloor. Piles are used to anchor each template in place. A simple analysis using linear theory was performed to aid in this initial TLP design cycle. The tendon's top and bottom connectors use a flex element that allows approximately a $\pm 10^0$ rotation with minimal bending moment [69]. The connector may be a complex mechanism and its selection is based on the desired installation method. As TLPs are designed for greater depths, the effect of environmental loads on the longer tendons and risers greatly increase and must be thoroughly analyzed and carefully designed [70].

The Auger TLP uses a spread well system (several individual wellheads) with a fixed drilling derrick. This design, in which the risers are spread over a larger area at the seafloor as compared to the deck, was implemented due to concerns of vortex shedding [71]. Rather than using a thruster system, a multi-leg catenary mooring system is used to meet the additional station keeping restoring force required to position the derrick over each wellhead. An eight-point catenary system is used on the Auger TLP. The components include mooring lines (5 *in.* diameter wire and 5 *in.* chain), submersible buoys, pile anchors and linear winches for positioning the hull. The addition of this system allowed for the reduction in the number and size of the tendons used, but with a high initial capital cost [67],[72],[73]. Yoshida et al. (1994) [74] discussed the active control of a TLP using a thruster system.

Bea et al. (1994) [75] have developed a methodology for comparing offshore production systems. It is based on analyzing the alternative designs through the project's complete life cycle. It is an interdisciplinary approach which results in a set of risk costs for each design. These values are an indication of where the greatest risks are and what cost-effective measures can be implemented to reduce the level of risk.

On Site Assembly

The construction of a TLP at sea presents many engineering challenges. The hull of the Auger TLP was towed 6,800 *miles* before being mated with its 24,000 *ton* deck during 10-15 *knot* winds [67]. The hull may be constructed as one unit and towed to the site or built in modules and assembled onsite [67],[68]. Each tendon consists of several steel pipes that may be welded onshore and towed to or assembled onsite. The former reduces the construction time at sea but the forces on the tendon during towing and up-ending must be considered. The Hutton and Snorre TLPs used threaded tendon couplings whereas the Auger TLP used snap together couplings, and single piece tendons were used in the construction of the Jolliet and Heidrun TLPs [67],[69]. Once all tendon connections are complete, the hull is deballasted and the tendons are pretensioned. Wybro (1995) [76] discusses current methods for tendon installation. A new method utilizing the Platform Arrestor Concept is presented to allow for the simultaneous lock-off of all the tendons. This offers reduced resonant behavior during installation and lower cost for deepwater operation.

Dynamic Response

The surge, sway and yaw resonant frequencies of TLPs are below those of the wave frequency range as defined by a power spectrum such as the Pierson–Moskowitz. The heave, pitch and roll resonant frequencies are above this range. The resulting frequency content of the response is a desirable feature of TLPs. Wind, waves and current will cause a TLP to oscillate about an offset position rather than its vertical position. This offset in the surge direction has a corresponding *set down*, the lowering of the TLP in the heave direction, which increases the buoyancy forces and results in a higher tension in the tendons than if it was in the vertical position. Higher order effects due to the nonlinear nature of the waves and nonlinear structural properties will affect the dynamic response and may be of interest. Papers that include varying levels of higher order effects are included in this review.

Rainey (1977) [77] presented a dynamic analysis of vertically–moored tethered buoyant platforms such as articulated towers and TLPs. The structure was modelled as a spherical mass representing the platform moored to the seafloor with a cable in zero viscosity water. The platform is treated as submerged in order to eliminate the effects at the free water surface. Heave, roll and pitch motions were considered to be of the second order and therefore not considered.

Enhancements to the above governing equation were then made to account for viscosity and free water surface effects. Small-wave theory was applied at the free water surface and added to the vertical load exerted on the platform by the waves. This results in the following governing equation for surge motion only:

$$x_{tt} + 2cx_t + \{1 + g(t)\}x = f(t), \tag{3.1}$$

where $f(t)$ and $g(t)$ represent the horizontal and vertical force histories respectively. This differential equation was then solved and a stability analysis was performed and compared to empirical results from a physical TLP model immersed in water of a depth of approximately 1.2 m. Tests were performed with both regular and irregular waves. Dynamic instability was found due to regular waves at the predicted wave periods and amplitudes. The irregular waves required a significantly higher amplitude to cause instability in the model.

Armenis et al. (1991) [78] performed a time domain dynamic analysis on a TLP. Non–linearities modelled included large tendon displacements, coupling between the hull and tendons, coupling between horizontal and vertical motions of the platform and the set down effect. Diffraction effects were included but second order wave forces due to potential effects were neglected.

Mekha et al. (1994) [79] studied the nonlinear effect of evaluating the wave forces on a TLP up to the wave free surface. Several approximate methods were evaluated for regular and irregular wave forces, with and without current, and compared to Stokes' second order wave theory. A two–dimensional TLP mathematical model consisting of four columns and four pontoons was created. The hull was modelled as a three degree of freedom rigid body undergoing surge, heave and pitch. The tendons were treated as massless springs providing axial and lateral stiffness at their connection with the hull.

Song and Kareem (1994) [80] analyzed a TLP as a platform–tendon coupled structure. Here, the tendon's curvature, mass and varying tension are modeled rather than approximated as an equivalent spring. The stochastic response to random wind and wave forces was determined. The response of the coupled system was then compared to a system in which the tendon was modelled as an equivalent spring. For surge, sway and heave motions the difference between mean value results was between 3% and 8%. The differences for the heave, roll and pitch motions were between 30% and 58%. These large differences were attributed to the heave, roll and pitch motions being controlled by the elastic force in the tendons and hence more sensitive to its characteristics. Use of an equivalent spring does not include the effect of roll and pitch on the tendon's stiffness. It was also shown that the importance of including the nonlinear effects of the tendon increases with greater water depth.

A parallel computation scheme to reduce computational time over that of a single microprocessor was also presented. The set of parallel microprocessors are divided into two groups, each solving one substructure in the time domain. At certain intervals data is shared between the two groups to couple the substructures.

Johnson (1994) [81] created a finite element analysis based software package to aid in the design of TLPs. A two-dimensional model with coupled hull and tendons was subjected to wave, wind and current loads. The hull consisted of a platform, four cylinders and four pontoons and was treated as a rigid body. The tendons were constructed of several beam–column elements and provide resistance to the hull's surge, heave and pitch motions. Linear wave theory and the Morison equation were used to determine the hydrodynamic loading on the structure. The finite element analysis may be performed with either the trapezoidal or Houbolt's method. The TLP response is presented graphically in real time.

Kim et al. (1994) [82] studied the nonlinear response of a TLP subjected to random waves, steady winds and currents. In the model, the hull and tendons are coupled and large displacements are allowed. The first and second order wave forces were determined using a higher-order boundary element method. The response was studied in the time domain utilizing a three-dimensional hybrid element method. The hull is treated as a rigid body and the tendons were modelled with three-dimensional beam elements connected at its top and bottom with spring-damper elements. Responses were analyzed for surge, heave and pitch motion, offset and setdown. The mean tension in the tendons was found to be about 15% higher than the pretension due to the large mean displacements. For a TLP model with 415 m tendons, the mean offset and setdown were found to be approximately 20 m and 0.5 m, respectively. Liu et al. (1995) [83] used the higher–order boundary element method to model a fixed and compliant TLP subject to second order mean and double–frequency wave loads. The second spatial derivatives of the first order velocity potential were included for the body and free surfaces. The tendons were modelled as massless springs. The free surface component was found to dominate the heave force and pitch moment. Large differences between the fixed and compliant TLPs were observed in the calculation of the yaw moment.

Kareem (1983) [84] presented time histories of wind velocity fluctuations as single and multiple–point Gaussian random processes. For single–point loading, an equation relating drag force as a function of fluctuating wind velocity was given for the surge direction. When multiple–point loading is desired, equations for force in the surge direction, and moments in the yaw and pitch directions were presented. Here, the wind velocity field is a function of time and its position in the sway–heave plane. In Kareem

(1985) [85] , these equations were transformed into the frequency domain. Both frequency and time domain analyses for a TLP subject only to wind were implemented to determine the response in the surge, yaw and pitch directions. The hull was modelled as a rigid body, and the tendons as massless springs. The motions were assumed to be uncoupled and solved numerically. Results showed that the surge mode is sensitive to the static and dynamic effects of wind forces. Pitch was minimally affected but may be of concern for fatigue analysis due to the low structural and hydrodynamic damping. Kareem recommended a frequency domain analysis at preliminary design stages and a time domain analysis at later stages.

In order to develop a better understanding of aerodynamic loading, mean aerodynamic force and moment coefficients were determined from wind tunnel tests [86] . These tests were conducted on a 1:128 scale model TLP hull with major components (for example, derricks, cranes, living quarters) in various configurations.

Wave induced forces for a TLP in an offset position may differ from those at the undisplaced position. Li and Kareem (1992) [87] analyzed this effect by adding displacement induced feedback forces to the wave induced loads calculated at the TLP's undisplaced position. Applying this method to a TLP model resulted in a surge response due to the feedback forces that were of the same magnitude as the wave frequency response. Drift induced by second order wave potential was not included.

Kareem and Li (1990,1993) [88] , [89] have studied the response of TLPs in the frequency domain. A procedure was described to analyze the coupled six degree of freedom motion of a TLP subjected to current and random wind and wave loads. Conventional frequency domain analysis was applied to linear and linearized systems, whereas nonlinear systems are typically solved utilizing a time domain approach. In this paper, the nonlinear terms are expanded into multivariate orthogonal polynomials. The zeroth–order terms in the polynomials represent the mean forces, the first–order corresponds to the wave, wind and damping forces, and the second-order accounts for the slowly varying drift forces. This may be solved with perturbation or iterative techniques with spectral convolution and quadratic transformation methods, but with significant increases in CPU requirements. A new spectral decomposition approach was developed to reduce computational requirements. It is a stochastic approach that more efficiently provides the input to be used by the perturbation or iterative methods. The resulting response closely matched that obtained from a time domain analysis with a reduction in CPU time.

In Kareem and Zhao (1994a) [90] , the surge response to wind, wave and current was found using a standard one degree of freedom governing equation. Wind and wave processes were assumed Gaussian, but the structural velocity was not. This led to highly nonlinear terms when expanded as an equivalent polynomial. In this study, terms above quadratic were neglected. A frequency domain analysis was performed with the response cumulants based on Volterra theory.

The response where only wind forces were included was determined in Kareem and Zhao (1994b) [91] . The response of a one degree of freedom TLP was presented in order to demonstrate the methodology. A nonlinear wind gust loading factor was also developed.

Li and Kareem (1993) [92] presented parametric models to generate time histories of waves. Autoregressive and moving averages (ARMA), convolution and interpolation techniques were used in these models. These methods were demonstrated with an application to a simple TLP modelled as a six degree of freedom rigid body subject to a random wave field. The dynamic response of the TLP was calculated in the time domain. In this paper, loading due to wave drift forces was not included, but was addressed in a subsequent report [93] . Li and Kareem (1995) [94] presented a stochastic decomposition technique to improve the efficiency of conventional frequency domain analysis and apply it to a TLP subject to random wind and wave fields.

The numerical response of a TLP subject to wind and wave loading was compared to the empirical results obtained by scaled tank testing by Vickery (1995) [95] . The numerical model was analyzed in both the time and frequency domains. The TLP's hull was modelled as a six degree of freedom rigid body moored to the seafloor with tendons that are assumed to remain straight. A boundary element model was created of the submerged portion of the TLP. As the hull translates in the surge direction, the tendon stretches, resulting in a higher tension. The total external force included first and second order wave forces, wind forces, and the drag term from the Morison equation. A fourth order Runge–Kutta differential equation solver was used to solve for the time–domain response. In the frequency domain analysis, nonlinearities were neglected.

A 1:200 scale experimental TLP model was tested in a wind wave flume by Vickery. The model consisted of a square deck, four columns and four pontoons. Each column was restrained with one cable representing the tendons. Additional modules were mounted on the platform including a scaled derrick and crew quarters. All six rigid body displacements of the hull, and one forward and one aft tendon tension were measured. The response was measured in a total of thirtyfive wind–wave conditions (seven wave heights each at five different wind speeds).

The comparison between the experimental, time and frequency domain analyses focussed on the surge response. Vickery concluded that the effect of wind on the surge response was highly dependent on the size of the incident waves. The importance of the response due to the wind diminished with an increase in wave height. For realistic combinations of wind and wave loads, the surge response showed an increase of approximately 30% to 100% as compared to a wave load only case. The heave response was also significantly influenced by the wind loads. The mean tension in the tendons was dominated by wind forces, but its dynamic tension was not affected.

Jefferys and Patel (1982) [96] created a linear analytical model, and both linear and nonlinear finite element models of a tendon. Assumptions included constant tension along the length of the tendon, pin joints at both ends, planar motion and negligible bending forces. Longitudinal motions were also neglected. This resulted in the following governing equation for a tendon:

$$my_{tt} + w(y_t, y_{tt}, r) = Ty_{zz}, \qquad (3.2)$$

where m is the mass per unit length of the tendon, y is the lateral displacement, r is the radius, T is the constant tension, z is the vertical coordinate, and $w(y_t, y_{tt}, r)$ includes the added mass and the drag force. In addition to solving Equation 3.2 for y, a modal analysis was performed and compared to the finite element models in the frequency and time domains. The finite element model where nonlinear drag forces were included was used in a study of the coupled dynamics of the hull with the tendons by Patel and Lynch (1983) [97] . This nonlinear drag force was included as an equivalent linear damping constant. Two approaches were presented to calculate this constant. The whole–tether approach assumes that the response is dominated by a solid body rocking mode and that higher modes are negligible. The element-by-element approach is an iterative technique that begins with an initial guess calculated using the whole–tether approach to determine the tendon's velocity and then includes this in subsequent iterations. The whole–tether method was found to be preferred. It is more computationally efficient and the differences between the two methods were negligible.

The hull was modelled as a rigid body with six degrees of freedom subject to wave forces. It was coupled to the finite element model of the tendon. This was accomplished by first calculating the response of the hull while assuming a quasistatic tendon stiffness. The hull's displacements were then used in the numerical analysis of the tendon's stiffness. The hull's response was then reevaluated. Only one iteration was performed since the hull's response was predominantly due to inertia and the effect of the tendon's stiffness was secondary.

Several perturbations were performed with varying physical properties. Patel and Lynch concluded that differences between the quasistatic and dynamic tendon models were minimal except for tendons of lengths on the order of 1500 m or greater. Therefore, the effect of the tendon's dynamics on the hull's response was only significant when the tendons were long, had a large mass per unit length and the hull's displacements were small. The bending stresses in the tendons were also determined to be small. The greatest values were found for short tendons with large outer diameters and thin wall thicknesses.

Patel and Park (1995) [98] studied the combined axial and lateral response of tendons. The tendon was modelled as a straight simply supported column subject to both horizontal and vertical motions at its top. This motion represented the response of the hull. The governing equation for the lateral motion of the tendon is:

$$My_{tt} + EIy_{xxxx} - (T_0 - S\cos\omega t)y_{xx} + B_v\,|y_t|\,y_t = 0, \qquad (3.3)$$

where $y(x,t)$ is the lateral displacement, M is the structural and added mass, EI is the structural flexural rigidity, T_0 is the constant axial tension, $S\cos\omega t$ is the axial force, and B_v is the drag induced viscous damping. At the tendon's top, the lateral force was assumed sinusoidal.

Assuming a solution for $y(x,t)$, the resulting equation of motion was solved numerically. The initial position of the top end was at the midpoint of the hull's surge motion. Results from studies of three different tendon lengths were presented. The analysis showed that a greater amplitude of vibration will result from the combined excitation than the individual axial and lateral excitations. This was particularly significant in even numbers of the instability region of the Mathieu stability chart. The frequency of oscillation was also found to be dependent on the relative magnitudes of the individual excitations for the combined response.

Lee and Lee (1993) [99] presented an analytical solution for a two–dimensional TLP subject to wave induced surge motion. It was approached as a boundary value problem divided into scattering and radiation problems. The surge response computed analytically compared favorably with the results of a numerical study that used the boundary element method.

Mullarkey and McNamara (1994) [100] studied a TLP subject to first-order wave forces. The effects of added mass and radiation damping were considered. Semi–analytical solutions for the hydrodynamics of the columns and pontoons were presented and compared to the results of a radiation–diffraction panel program for wave–body interactions.

Park et al.(1991) [101] performed a reliability analysis on four and six column TLPs. Two failure criteria were considered for the tendons at the four corners: ultimate tensile strength and negative tension. Negative tension, or slacking in the tendon, may result in high impulsive stresses at the return of positive tendon tension. The TLP was modelled as a six degree-of-freedom system with the hull treated as a rigid body and the tendons as massless springs. It was subject to a severe storm whose intensity was represented by a maximum wave height treated as a random variable. Only first-order wave effects were considered. Two angles of wave attack were considered, zero degrees and approximately 45^0. The failure probabilities were determined for a 20 *year* service life with the 45^0 case being more critical. The results indicated that negative tension was the governing failure criterion. Although not modelled, a sensitivity analysis was performed to indirectly study the effect of wind, current and second-order wave forces. These forces may cause horizontal mean drifts. Inclusion of drift in the model was found to increase the tension in the tendons, thereby improving reliability by reducing the chance of tendon slacking.

Seismic Response

Liou et al. (1988) [102] , Kawanishi et al. (1991) [103] , Kawanishi et al. (1994) [104] and Venkataramana (1994) [105] studied the response of TLPs to vertical seismic excitation. If sufficiently large, the vertical displacements may cause slackening in the tendons. Horizontal displacements were on the order of 10^2 times greater than the vertical displacements. It was also noted that the horizontal and vertical displacements were reduced when the pile-soil foundation was included in the dynamic model as compared to a fixed base.

Springing and Ringing

Springing is a steady state response due to first and second-order wave diffraction sum–frequency effects. Ringing is a transient response due to various nonlinear free surface effects. Springing and ringing may contribute to fatigue and cause jerks in the platform that may affect the comfort of the crew. Natvig and Teigen (1993) [64] , in their review of hydrodynamic challenges in TLP design, included the need for further research of springing and ringing. The designers of the Heidrun Concrete TLP paid special attention to the effect of ringing [106] . The following papers discuss current investigations into these effects.

Second-order wave forces are quadratically nonlinear and may cause low-frequency drift oscillations and affect the high-frequency vertical modes. This nonlinearity is due to the wave forces varying with the square of the incident wave heights, the variable wetting of the TLP columns, and the effect of the velocity squared term in the Bernoulli equation [107] , [108] , [109] . Solving the quadratic velocity potential will result in pairs of frequencies. Adding or subtracting one frequency from another results in a frequency which is referred to as a sum– or difference–frequency.

Neglecting these second–order wave forces, the TLP's surge, sway and yaw resonance frequencies are below that of the wave frequency range. The heave, pitch and roll resonance frequencies are above this range. If the second–order wave forces are included, the difference–frequencies may be in a bandwidth to increase the TLP's horizontal response, and the sum-frequencies may increase its vertical response. The additional horizontal response could affect drilling operations and the sum–frequencies will create a stretching of the tendons resulting in a springing effect. Springing can lead to earlier fatigue in the tendons.

Kim et al. (1991) [107] studied the TLP responses under random sea conditions. A digital second-order polyspectral analysis was used to compute a power transfer function. The results of an experimental 1:54 scale TLP model was used to quantify the amount of power extracted from the incident wave spectrum and transferred to each spectral component of the sum- and difference-frequency response. In the linear spectral model, the input wave only affected the surge response at the same frequency. For the second–order spectral model, the sum– and difference–frequencies were important and considered.

Surge response and wave height were measured from the scaled model simulating a TLP moored in $1500\,ft$ of water. The TLP was subject to 45^0 unidirectional irregular waves generated according to the Pierson–Moskowitz spectrum. The results showed that the low-frequency surge response is dependent on the difference–frequencies with a far smaller contribution from the sum–frequencies.

Marthinsen and Muren (1991) [110] analyzed the measured response from the Snorre TLP. This TLP operates in the North Sea at a water depth of $310\,m$ and produces 190,000 barrels of oil each day. The springing standard deviations were presented for the measured roll and pitch response and compared to the calculated pitch response. These calculations were performed with second order potential theory and were found to underestimate the response.

Natvig and Vogel (1991) [108] explored the effect of sum–frequencies on tendon fatigue and extreme response. They studied a TLP subject to direct wave, slowly varying wave drift, wind gust and sum–frequency excitations. A linearized frequency domain procedure was implemented and the effect of various types of damping was presented. The second–order wave sum–frequency response analysis was based on Quadratic Transfer Functions (QTF) obtained from three different institutions. The QTF represents the total wave force normalized with respect to the incoming wave amplitudes. Only sum–frequencies near the natural frequencies for motion in the vertical plane were considered. In cases where the QTF data did not include frequencies near these resonant frequencies a linear interpolation was performed. The greatest force spectral density was found where a sum–frequency had a corresponding difference–frequency which was smallest.

The influence of sum–frequency forcing was highly dependent on the damping in the system. Potential damping due to wave radiation, viscous damping, column footing damping from model tests, soil damping and structural damping were discussed. Structural damping, though normally neglected, should be considered. For typical welded structures 0.5% to 0.8% can be used, and for the TLP components a value of 0.2% was recommended.

A comparison was made between the three QTFs for one sum–frequency and a single wave heading. The vertical sum-frequency forcing was found to be less important than the moment forcing. Problems can also result from using a single heading since the QTF data set may contain a lower or higher energy level at that heading. This will greatly affect tendon fatigue analysis. The second–order sum–frequency forcing also had a significant impact on the extreme tendon response undergoing a large 100–year sea state.

Natvig (1994) [111] studied the ringing response observed in TLPs. Springing, due to first and second–order wave diffraction sum–frequency effects, has a slowly varying amplitude with only moderate variation. A TLP's response due to ringing resembles that of a struck bell. Its amplitude builds quickly and then decays slowly. It is a transient state that can only occur when the steady state springing is present; however, springing can occur without ringing. It is believed that ringing type responses are due to the various nonlinear effects of free surface variable wetting. The maximum ringing force may be as high as the forces due to other dynamic effects. However, since ringing rarely occurs, it may not have a significant effect on fatigue.

The response from model tests was presented. Ringing was found for all headings with diagonal waves being the most significant. Ringing was considerably reduced with slight changes in heading from the diagonal. Single high waves with steep fronts and/or backs, but not breaking waves resulted in ringing. Ringing was not sensitive to increases in damping but was affected by vertical shifts in the TLP's center of gravity.

An analytical model was also created where the tendons were treated as massless springs. Certain nonlinear effects were included and the response for one loading case was compared to that of a model test. The importance of each effect in this loading case was presented. The frequency dependent wave force coefficients from diffraction, the frequency dependent wave crest amplification around a column, the time derivative of the added mass, and the wave slap on the columns were found to be of major importance.

Gurley and Kareem (1995) [112] simulated the ringing response using a nonlinear Duffing oscillator. The pitch response of a 10 m long immersed column subjected to linear and nonlinear waves was studied. The Morison equation and stretching theory was used to compute the wave force up to the instantaneous water level. (Mekha et al. (1994) [79] reviewed above compared stretching theory with other approximations.) The effect of the change in added mass due to the varying water level was included. Ringing was only present when nonlinear wave effects were included. Increasing the system center of mass eliminated the ringing response and the effect due to changes in the level of damping was insignificant. The response due to the grouping of waves was also studied. Ringing was observed for a single group of large waves, with the effect lessened if multiples groups were applied. These results compared favorably with that of Natvig (1994).

Effect of columns

Columns were often incorporated into the single rigid body representing the hull in the papers above. The columns are the structural elements that pierce the water surface. Therefore, in addition to wind and current forces, they are subject to wave loads that include the wave free surface effects. The wake created by one column may affect another making the positioning of the columns important. The columns are also subject to variable added mass and damping.

Chung (1994) [113] studied the effect on the added mass and wave damping coefficients due to the free surface effect. Experiments showed a larger effect on the added mass coefficient for the vertical oscillation than for the horizontal oscillation. Others have studied hydrodynamic forces on vertical cylinders either numerically [114] or experimentally [115], [116], [117].

Vortex shedding and vortex induced vibration can be of significance in the study of TLPs since it can result in large amplitudes of vibration [58]. This effect has been investigated numerically [118] and experimentally [119], [120], [121], [122]. Chung et al. (1994) [123] discussed means to reduce the effect of vortex shedding that included the use of helical strakes (cables wrapped around the cylinder along a helical path).

Niedzwecki and Huston (1992) [124] conducted 1:250 scaled wave tank tests. A four column TLP model with and without pontoons was studied with varying column spacing. Regression formulas were developed for the wave run-up on a forward leg as a function of horizontal wave velocity and incident wave height. Ratios of wave run-up between the aft and forward columns were presented for different column spacing as a function of wave period. Wave run-up was found to be greater for closer column spacing. Wave upwelling, the local amplification of the incident wave field, can damage the underside of the platform if sufficiently large. Raising the platform increases cost and decreases stability. Maximum upwelling values as a function of incident wave period were presented. As the column spacing increased, the magnitude of the wave upwelling decreased and the wave period with the peak amplitude shifted to lower wave periods.

Niedzwecki and Rijken (1994) [125] conducted experimental studies to analyze the effect of nonlinear wave runup on TLPs and the nonlinear response of risers. The hull was modelled as a single truncated cylinder piercing the water surface. A random wave field was used and statistical moments and probability plots were presented. The surface elevation of the waves was found to be non–Gaussian with higher crest amplitudes and lower trough amplitudes. Increased wave steepness led to a more pronounced non–Gaussian behavior. Testing performed with the truncated cylinder in place resulted in wave runup amplitudes which significantly exceeded the significant wave height.

Duggal and Niedzwecki (1995) [126] modelled a TLP tendon or riser as a single flexible cylinder to simulate a depth of $1000\ m$. The cylinder was mounted in a wave basin and pretensioned. A different curvature response pattern was measured just below the still water level as compared to the middle level of the cylinder. This was attributed to the depth dependency of the Keulegan–Carpenter number. The probability density function of the measured curvature was found to be non–Gaussian and would result in a larger curvature than a Gaussian assumption would predict.

Faltinsen et al. (1995) [127] analyzed a vertical circular cylinder subject to nonlinear wave loads. The authors found that the second and third order wave forces near the free surface were of comparable magnitude. Kim (1993) [128] presented equations in closed form to determine the interaction of waves between multiple vertical circular cylinders. Wavelength, column spacing and wave heading were determined to be important factors. Zhu et al. (1995) [129] studied the vibration of a circular cylinder in the wake of another circular cylinder. Two cylinders were positioned at six different distances apart. Unsteady–flow theory and direct measurements were used. It was found that the position of the cylinder in the wake significantly affected its stability. Mizutani et al. (1994) [130] analyzed wave forces on two and three large diameter cylinders. They found that wave diffraction affected the lead cylinder more in the two cylinder experiment and the center cylinder in the three cylinder study.

Tendons

In many of the reviewed papers, tendons were treated as massless springs. In the more advanced structural models, the tendons themselves were modelled. Tendons can be modelled as beams, cables or strings with varying degrees of complexity. Tendons and risers have many similarities in that they are both very long submerged hollow pipes subject to hydrodynamic loads. As the water depth increases, the need to consider higher order effects in both the loading and the structure increases. Two fundamental differences between the two is the extremely high tension experienced by the tendons and the fluid flowing within the riser [131] . Analysis techniques for risers may be applied to tendons. Patel and Seyed's (1995) [132] review paper includes seventy-four references on modeling and analysis techniques for flexible risers.

As with the columns discussed above, vortex shedding and vortex induced vibration may need to be considered. Although the tendon diameters are significantly smaller than the column diameters, there are generally multiple tendons at each corner of the hull. This may lead to interactions between the tendons at one corner as well as between the corners. Bokaian (1994) [133] modelled a tendon or riser subject to vortex shedding in a vertically sheared flow. For a tendon of uniform mass and cross-section, a closed form solution was presented. It showed that multiple structural modes are usually excited by the shear flow. The numerical results were in agreement with reported experimental observations.

Several papers are cited here that present methods to aid in modeling tendons as beams: [134] , [135] , [136] , [137] , [138] , and [139] . The dynamic response of cables may also be useful in the modeling of tendons; [140] , [141] , [142] , [143] , [144] , [145] , [146] , [147] , [148] , [149] , [150] , including the cases of extreme tension [151] , and negative tension [152] .

Patel and Park (1995) [153] investigated tendon response subject to short duration tension loss. The governing equation of motion for the lateral motion of the tendon was solved analytically to study the dynamic pulse buckling behavior. Since higher modes are affected by pulse bucking, bending stiffness was included. Compressive stress as a function of time was presented. It was found that TLPs can be designed to allow for a tension loss of a few seconds.

Mekha et al. (1996) [154] created three different tendon models. The first was as a massless spring to provide a constant lateral stiffness with TLP setdown neglected. The second allowed for time varying axial forces in the spring with one third of the tendon mass lumped at the attachment points. The third tendon model consisted of flexural beam elements and allowed for the time varying axial forces. Hydrodynamic loads on the tendons were included in the third model only. Wheeler's approximation was used to determine the wave forces up to the wave free surface. This stretching method extends the wave kinematics in the same form as calculated to the mean water level, a hyperbolic cosine function, up to the free surface. The hull was modelled as a rigid body subject to regular waves and analyzed at different wave depths, wave heights and wave periods.

The amplitude of the surge response was found to be independent of the model used. For a constant wave period it increased linearly with wave height, and decreased linearly with increasing wave frequency for a constant wave height. The surge mean offset may more than double for the third model where hydrodynamic loads on the tendons were included. The maximum tendon forces were not affected by the hydrodynamic loads. However the minimum decreased by ten percent. Neglecting the inclusion of the hull setdown in the first model resulted in a maximum tendon force that was about fifteen percent higher.

Equation of Motion

The equation of motion for the TLP is obtained directly from equations (1.166) and (1.167), by setting the internal fluid properties to zero, that is, set density, velocity, acceleration and internal pressure to zero, $\rho_{fl} = 0, v_{fl} = 0, a_{fl} = 0, p = 0$,

$$
\begin{aligned}
&[A_T (\rho_T + C_A\rho)] (y_{tt} + u_{tt}) + [A_T (\rho_T + C_A\rho) w_t] y_{xt} + \\
&[A_T C_A\rho w_{tt} + g (A_{fl} (\rho - \rho_{fl}) + A_T (\rho - \rho_T))] y_x + EI y_{xxxx} + \\
&[-\rho g \left((A_{fl} + A_T) (l - x - w) + \frac{1}{\cos\varphi} A_b (d - L) \right) + C_v y_t - I_T y_{xxtt} + \\
&(M + (\rho_T A_T + C_A\rho) (w + l - x)) (g + w_{tt}) - A_T (\rho_T + C_A\rho) w_t^2] y_{xx} = Q,
\end{aligned}
\tag{3.4}
$$

and

$$
\begin{aligned}
&-\frac{1}{2} M L_b \left[\left(w_{tt} - \frac{1}{2} (y_{tt}(L) y_x(L) + 2 y_t(L) y_{xt}(L) + y(L) y_{xtt}(L)) \right) \sin\varphi + \\
&\left(w_t - \frac{1}{2} (y_t(L) y_x(L) + y(L) y_{xt}(L)) \right) \varphi_t \cos\varphi - \dot{V}_T^y \cos\varphi + V_T^y \varphi_t \sin\varphi \right] + \\
&J_p \varphi_{tt} = M_{gb} + M_{wind} + M_{wave} - M_b.
\end{aligned}
\tag{3.5}
$$

These equations are accompanied by the following boundary conditions,

Pinned:

$$
x = L + w; \qquad \begin{cases} S + F_y^R \cos\theta - F_x^R \sin\theta = 0 \\ M_b = 0 \end{cases}
\tag{3.6}
$$

$$x = w; \quad \left\{ \begin{array}{l} y = 0 \\ M_b = 0 \end{array} \right. \tag{3.7}$$

Fixed:

$$x = L + w; \quad \left\{ \begin{array}{l} M_b + M^R = 0 \\ S + F_y^R \cos\theta - F_x^R \sin\theta = 0 \end{array} \right. \tag{3.8}$$

$$x = w; \quad \left\{ \begin{array}{l} y = 0 \\ y_x = 0, \end{array} \right. \tag{3.9}$$

where all the symbols have been defined in Chapter 2.

Next the equations are solved numerically to explore the dynamic response characteristics of representative platforms under loading conditions that demonstrate possible behavior.

Deterministic TLP Response

In this section the response of the tension leg platform for different environmental conditions and different physical parameters is investigated. Also, the response to various boundary conditions are compared. The following aspects of the structural response are investigated:

- *Pinned–pinned* boundary conditions are compared to *fixed–pinned*.
- The fundamental frequency of oscillation along with the equilibrium position with no external loading are established.
- Response to wind and current excitation is calculated.
- The structural response to wave excitation is found.
- The response to base excitation is examined.

The physical parameters used in the analysis are shown in Table 3.4. Unless otherwise mentioned, *pinned–pinned* boundary conditions are assumed. These parameter values are chosen to be representative of those found in practice.

l	d	M	D_b	D_o	D_i
$100\ m$	$130\ m$	$15.0 \times 10^6\ Kg$	$20\ m$	$1.0\ m$	$0.95\ m$

3.4 Physical Parameters for the TLP.

Boundary Conditions

The cable connecting the deck to the sea floor is analyzed using two different sets of boundary conditions at the sea floor: *fixed* or *pinned*. Due to the very small ratio between the cable diameter and its length, $D_0/l = 0.01$, it acts more like a cable than a beam and has a negligible bending rigidity. Thus, one expects to find almost no differences between *pinned–pinned* and *fixed–pinned* responses. Fig. 3.2 shows a comparison of the free vibration responses for the two cases in the time and frequency domains. No difference can be seen between the cross–plots except at the higher frequencies.

From Fig. 3.3, which shows the responses for both cases, due to wave, current

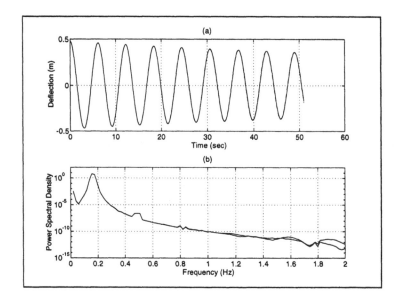

3.2 Comparison of the Free Vibration Response, $pinned - pinned$ vs. $fixed - pinned$. (a) Time Domain, (b) Frequency Domain.

and wind, it is clearly seen that a cable approximation (no bending rigidity) is quite accurate.

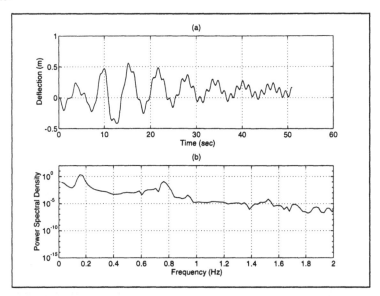

3.3 Comparison of the Response when Subjected to Wave, Current and Wind. (a) Time Domain, (b) Frequency Domain.

Fundamental Frequency

In this section, the free vibration of the TLP is investigated. Fig. 3.4 shows the response of the top end of the TLP in the time and frequency domains, when it is subjected to nonzero initial conditions. From Fig. 3.4 (a), it is clearly seen that the response decay in time is due to the drag force. From the frequency domain response of Fig. 3.4 (b), the natural fundamental frequency is found to be $\omega_n = 0.18\ Hz$. The odd harmonics of the fundamental frequency are due to the asymmetric nature of the drag force.

Fig. 3.5 shows the TLP at its initial position (solid line), its position at $t = 20$ s (dotted line), and at $t = 50\ s$ (dashed line). It is seen that the deck's motion is practically lateral, since the moments acting on the deck are negligible. Finally, Fig. 3.6 depicts the total response of the TLP.

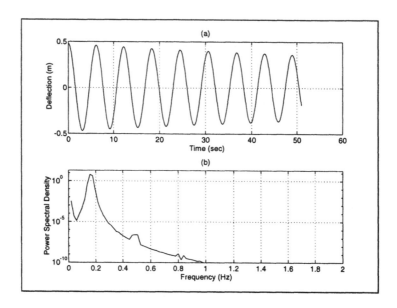

3.4 Free Vibration Response, (a) Time Domain, (b) Frequency Domain.

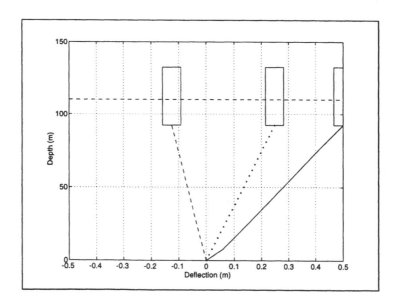

3.5 TLP's Position at $t = 0\,s$ (solid line), $t = 20\,s$ (dotted line), and $t = 50\,s$ (dashed line). The Initial state is given by the solid line.

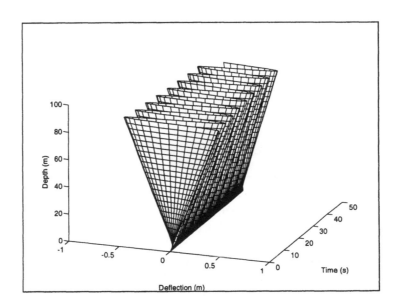

3.6 Total Response of the TLP to Nonzero Initial Conditions.

Response to Wind and Current

In this section the modified equilibrium position due to the presence of wind and current is presented. Fig. 3.7 depicts the equilibrium position of the TLP due to a wind speed of $u_{wind} = 30$ *m/s*. The position of the deck is clearly seen to be tilted at an angle of $\varphi \cong 0.1$ *rad* due to the wind force.

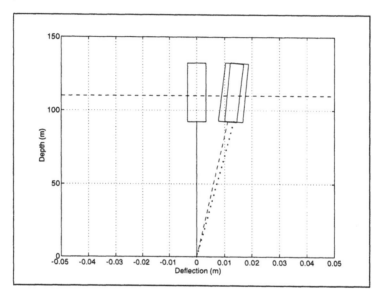

3.7 Equilibrium Position Due to Wind Velocity of $u_{wind} = 30$ *m/s*, at $t = 0$ *s* (solid line), $t = 20$ *s* (dotted line), and $t = 50$ *s* (dashed line).

Fig. 3.8 shows the time and frequency domain responses of the top end of the TLP. The response decays due to the air drag force and oscillates about a nonzero position. From the frequency domain figure, two significant frequencies are clearly seen. One is the first fundamental frequency of the cable, $\omega_n = 0.18$ *Hz*, and the other is the natural frequency of rotation of the deck, $\Omega_n = 0.02$ *Hz*.

In the presence of a current of speed $U_c = 2$ *m/s*, both the deck and the cable are subjected to drag force. Fig. 3.9 shows the position of the TLP at its initial position at $t = 0$ *s* (solid line), its position at $t = 20$ *s* (dotted line), and at $t = 50$ *s* (dashed line). The structure reaches an equilibrium position which is proportional to the current velocity squared.

The time and frequency domain responses of the top end of the cable are depicted in Fig. 3.10. The response decays faster under these conditions than when subjected to wind since the drag (damping) force is proportional to the density of the medium, and the density of water is 1000 times larger than that of air. The total response in the presence of current is depicted in Fig. 3.11. The decay due to damping is clearly seen.

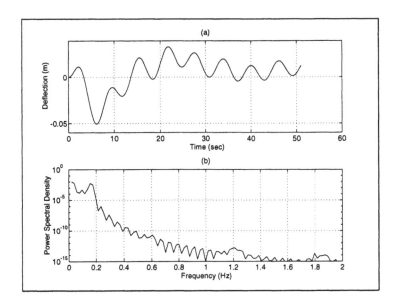

3.8 Response to Wind Velocity of $u_{wind} = 30\ m/s$, (a) Time Domain, (b) Frequency Domain.

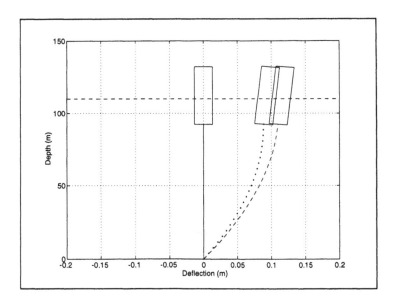

3.9 Equilibrium Position in the Presence of Current $U_c = 2\ m/s$, at $t = 0\ s$ (solid line), $t = 20\ s$ (dotted line), $t = 50\ s$ (dashed line).

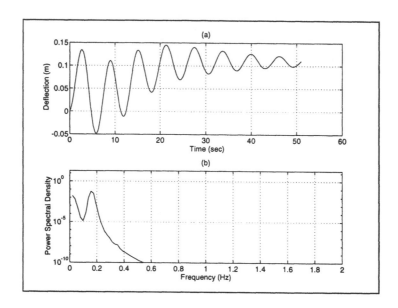

3.10 Equilibrium Position in the Presence of Current Velocity of $U_c = 2$ *m/s,* (a) Time Domain, (b) Frequency Domain.

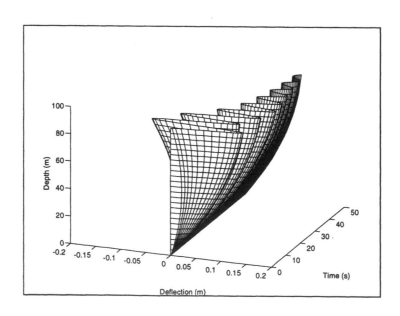

3.11 Total Response in the Presence of Current Velocity of $U_c = 2$ *m/s.*

Response to Wave Excitation

In this section the response to wave excitation is presented. Fig. 3.12 depicts the position of the TLP at three different times, $t = 0\ s$ (solid line), $t = 20\ s$ (dotted line), and $t = 50\ s$ (dashed line), while subjected to a wave height of $H = 2\ m$ and frequency $\omega_{wave} = 0.25\ Hz$. From the figure it is seen that the rotation of the deck is practically negligible at $\varphi = 0.05\ rad$.

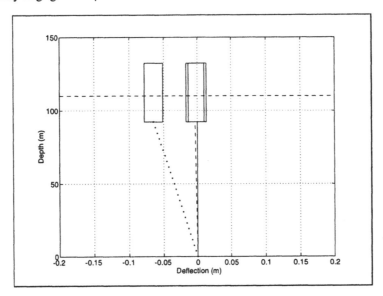

3.12 Response to Wave Excitation, $H = 2\ m$, $\omega_{wave} = 0.25\ Hz$ at $t = 0\ s$ (solid line), and its position at $t = 20\ s$ (dotted line) and $t = 50\ s$ (dashed line).

From the time and frequency domain responses of the end of the cable shown in Fig. 3.13, it is seen that the cable oscillates about a zero displacement, which is the equilibrium position in the absence of wind and/or current. The fundamental frequency $\omega_n = 0.18\ Hz$ is clearly observed but the wave frequency $\omega_{wave} = 0.25\ Hz$ is hardly seen due to the low amplitude of excitation. The total response is depicted in Fig. 3.14.

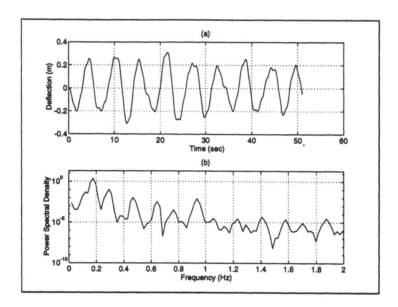

3.13 Response of Cable End to Wave Excitation $H = 2\,m$, $\omega_{wave} = 0.25\,Hz$, (a) Time Domain, (b) Frequency Domain.

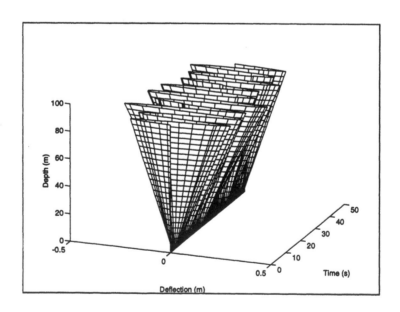

3.14 Total Response to Wave Excitation with $H = 2\,m$ and $\omega_{wave} = 0.25\,Hz$.

When waves, current and wind are present concurrently, the response oscillates about a nonzero equilibrium position, as can be clearly seen from Fig. 3.15. These figures depict the time and frequency domain responses of the top end of the cable when it is subjected to a wave with $H = 2\ m$ and frequency $\omega_{wave} = 0.25\ Hz$, a current of $U_c = 2\ m/s$, and a wind of $u_{wind} = 30\ m/s$.

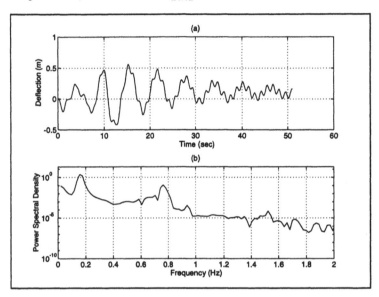

3.15 Response to Wave, Current and Wind Excitation. (a) Time Domain, (b) Frequency Domain.

Fig. 3.16 shows the TLP's position at three different times, $t = 0\ s$ (solid line), $t = 20\ s$ (dotted line) and $t = 50\ s$ (dashed line). From the figure it is seen that the deck is tilted, and the maximum angle that is detected is $\varphi = 0.16\ rad$. Fig. 3.17 depicts the total response of the TLP due to wave, current and wind.

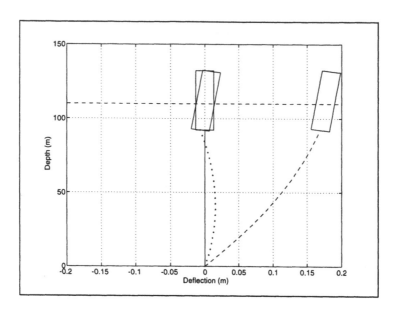

3.16 Response of Cable End to Wave, Current and Wind Excitation at $t = 0\ s$ (solid line), and $t = 20\ s$ (dotted line) and $t = 50\ s$ (dashed line).

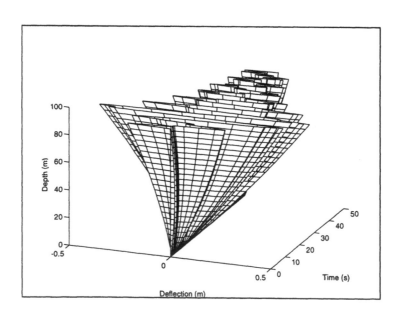

3.17 Total Response to Wave, Current and Wind Excitation.

When the wave frequency is near the fundamental frequency of the cable or its multiples, $\omega_{wave} \cong \omega_n$ or $\omega_{wave} \cong 2\omega_n$, the response behaves as though in a resonance. Fig. 3.18 shows the response of the cable end in the time and frequency domains with $H = 2\,m$ and $\omega_{wave} \cong \omega_n$. The response beats with relatively high amplitudes. From the frequency response, two close peaks are seen; one is the natural frequency and the second is the wave excitation frequency. The total harmonic response is depicted in Fig. 3.19.

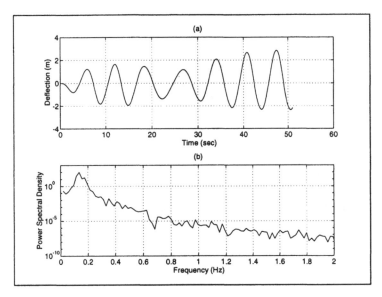

3.18 Harmonic Response of Cable End with $H = 2\,m$ and $\omega_{wave} \cong \omega_n$. (a) Time Domain, (b) Frequency Domain.

Subharmonic behavior for $\omega_{wave} \cong 2\omega_n$ is shown next. The beating phenomenon occurs here as well, as shown in Fig. 3.20 and 3.21.

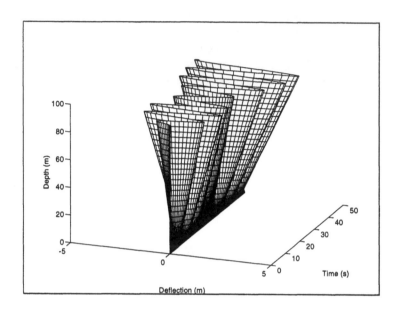

3.19 Total Harmonic Response with $H = 2\ m$ and $\omega_{wave} \cong \omega_n$.

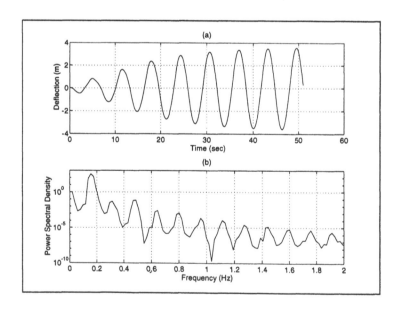

3.20 Subharmonic Response of Cable End for $H = 2\ m$ and $\omega_{wave} \cong \omega_n$. (a) Time Domain, (b) Frequency Domain.

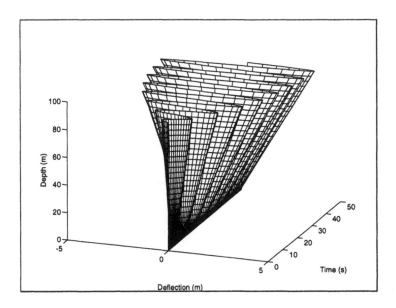

3.21 Total Subharmonic Response with $H = 2\,m$ and $\omega_{wave} \cong 2\omega_n$.

Response to Base Excitation

Base excitation can be used to simulate earthquake motion. The base is assumed to oscillate harmonically as follows,

$$w \;=\; 0.1\cos\omega_b t \qquad\qquad (3.10)$$

$$u \;=\; 0.2\cos\omega_b t, \qquad\qquad (3.11)$$

where $\omega_b \cong \omega_n$. More complex and realistic ground motions can be introduced, as shown previously for the articulated tower. In this case, the response of the cable is unstable as can be seen from the time domain response in Fig. 3.22. Fig. 3.23 captures this instability in the total response of the TLP to base excitation.

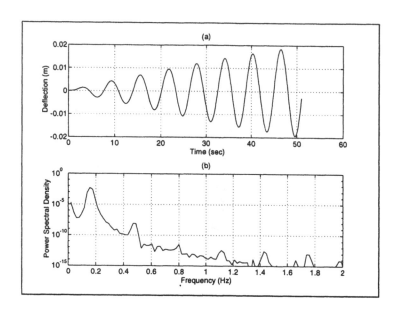

3.22 Response of Cable End to Base Excitation. (a) Time Domain, (b) Frequency
Domain.

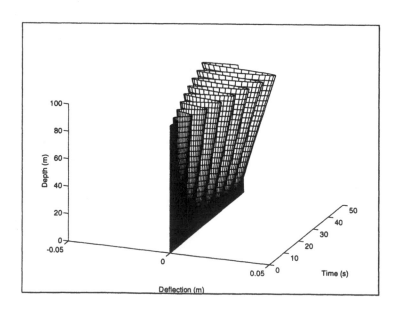

3.23 Total Response to Base Excitation.

Summary

The various primary loading cases for a TLP have been examined, including excitation due to waves, current, wind, as well as possible base loading due to seismic activity. Our purpose has been to demonstrate possibilities rather than actual design cases, or to be exhaustive in our parameter variations. Such models as developed and used for simulations can be very valuable as part of a preliminary analysis and for the initial sizing of the components to be used in a full design, which would then be verified using scale models testing and finite element techniques.

4
Risers with Flowing Fluid

In this chapter, the problem of the *riser* is addressed and a generalized approach is taken to develop the governing equations of motion for a riser type structure. The intention is to use the term *riser* in its most general sense. Flexible risers (and pipes) have found a broad range of applications, from pressure vessels to offshore production and transport facilities. Such applications are described in a large body of literature that is only briefly introduced here. The review paper by Patel and Seyed [98] provides a thorough introduction to some of the key literature. The question of the modeling of risers with internal flow for various boundary conditions and diverse application is a difficult one. Physically, this is a complex *fluid–structure interaction* problem where the structure undergoes large deflection while vibrating nonlinearity, and the fluid may be generally turbulent and accelerating. Technically, one has cost and engineering issues, with the latter including topics such as sealing materials, connections, and questions of insulating properties, among many others. The purpose here is to develop a *reasonable analytical model* for the nonlinear response of fluid conveying risers. Of course, assumptions are necessary and numerical simulations are required. But in the complex engineering analysis and design world, considerations such as those developed and presented here are important prerequisites to the eventual large–scale computational models that precede actual design. The models offered here permit the analyst and designer to gather the larger picture of possible dynamic regimes of behavior and then to begin to *size* the actual structure that is to be build.

Several papers that have provided overviews on this subject are listed next. In addition to the above review paper, Triantafyllou [155] reviews some basic aspects of cable and chain dynamics, and although such structures may be viewed as of a different class than pipes and risers, one can learn much from their dynamic analysis. Similarly, much has been learned from the last chapter on tension leg structures that can be applied here. Seyed and Patel [153] presented a short overview of the present status of research on flexible risers. They describe the prominent features of various analytical and numerical models in use, and set out to investigate the sensitivities of flexible riser performance to a series of structural and environmental parameters. Moe, Stromsem and Fylling [156] discuss the effects of axial tension on the dynamics of fluid conveying pipes. The most common boundary conditions are discussed. Sometimes the vibration amplitudes are unacceptable. Discussing the absorption of such energies are Aso, Kan, Doki and Iwato [157]. They show how a specific vibration absorber reduced resonance amplitudes by over 40%. Finally, McCone [158] discusses some of the technical challenges of the design of flexible pipes.

This chapter focuses on the specific problem of modeling the *large amplitude elastic vibration of a fluid conveying pipe*, with an eye toward characterizing its dynamic behavior. The analytical/numerical model developed is of a general nature and can be specialized further for application to other technical problems. In addition, this framework can be used to generate a broad range of parametric studies, although only representative results are included here.

Equation of Motion

The equations of motion for the riser are obtained directly from the general formulation of Chapter 2, equations (1.166) and (1.167). Assuming that the top end of the riser is attached to a deck that has very large inertia compared to the inertia of the riser, so that the motion of the deck is not effected by that of the riser, equation (1.167) is not needed. This results in the equation of motion for the riser,

$$
\begin{aligned}
&[A_T \left(\rho_T + C_A \rho\right) + \rho_{fl} A_{fl}] \left(y_{tt} + u_{tt}\right) + \\
&[2\,\rho_{fl} A_{fl} \left(w_t + v_{fl}\right) + A_T \left(\rho_T + C_A \rho\right) w_t] y_{xt} + \\
&[\rho_{fl} A_{fl} \left(w_{tt} + a_{fl}\right) + A_T C_A \rho w_{tt} + g \left(A_{fl} \left(\rho - \rho_{fl}\right) + A_T \left(\rho - \rho_T\right)\right)] y_x + \\
&C_v y_t + [p A_{fl} - \rho g \left(\left(A_{fl} + A_T\right) \left(l - x - w\right) + A_b \left(d - L\right)\right) + \\
&\frac{3}{2} \rho_{fl} A_{fl} \left(2\,w_t + v_{fl}\right) w_t - I_T y_{xxtt} + EI y_{xxxx} + \\
&\left(M + \left(\rho_T A_T + \rho_{fl} A_{fl} + C_A \rho\right) \left(w + l - x\right)\right) \left(g + w_{tt}\right) - \\
&A_T \left(\rho_T + C_A \rho\right) w_t^2] y_{xx} = Q
\end{aligned}
\tag{4.1}
$$

where as a reminder, l is the pipe's length, I_T is the mass moment of inertia of a pipe element and I is the cross section moment of inertia of a pipe element.

This equation is accompanied by the following boundary conditions. It is assumed that the riser is pinned at the top and at the bottom. The top end of the riser is assumed to follow the motion of the deck, which can be represented as a harmonic function of time. Thus, the boundary conditions are

$$
x \;=\; L + w; \quad \begin{cases} y = y_0 \cos \omega_t t \\ M_b = 0, \end{cases}
\tag{4.2}
$$

$$
x \;=\; w; \quad \begin{cases} y = 0 \\ M_b = 0, \end{cases}
\tag{4.3}
$$

where y_0 is the displacement magnitude at the top and ω_y is the oscillation frequency at the top.

In the analysis, the internal fluid velocity and acceleration, v_{fl} and a_{fl}, are assumed to be time dependent as follows,

$$v_{fl} = v_0 + v_1 t + v_2 \cos \omega_{fl} t \qquad (4.4)$$

$$a_{fl} = v_1 - v_2 \omega_{fl} \sin \omega_{fl} t. \qquad (4.5)$$

Each term is analyzed separately. The first term, v_0, represents the constant velocity of a fully developed flow. The second term, $v_1 t$, represents a transient state where there exists a constant acceleration and the third term is representative of a pump that creates an oscillatory flow. v_0, v_1 and v_2, as well as ω_{fl}, can be specified for a particular application. The purpose here is to demonstrate a few possibilities.

Deterministic Riser Response

In this section, the response of the riser to different environmental conditions and different physical parameters is investigated. Also, response in the presence and absence of internal fluid flow is compared. Thus, it may be possible that under certain conditions and for certain vibration cases, the fluid can be neglected. In general though, this cannot be assumed *a priori*. The following aspects of the structural response are investigated here:

- The fundamental frequency and equilibrium position without external loading is evaluated.
- The response to wave and current excitation is simulated.
- The response to base and top excitation is added to the previous analysis.
- The significance of an accelerating internal flow is assessed.

Table 4.5 shows the physical parameters used in the analysis;

l	d	M	D_o	D_i	v_{fl}
$100\,m$	$110\,m$	$20.0 \times 10^6\,Kg$	$0.21\,m$	$0.2\,m$	$300\,m/s$

4.5 Physical Parameters for the Riser.

These parameter values are used as representative of the scales of actual riser problems encountered in practice. As the purpose of this work is not to become a design handbook of cases, other parameter values have not been included here.

Fundamental Frequency

In this section, the free vibration of the riser is investigated. Fig. 4.1 shows the response of the top end of the riser in the time and frequency domains, when subjected to nonzero initial conditions. From Fig. 4.1(a) it is clearly seen that in the absence of internal flow (dashed line) the response decays with time due to the drag force, while the response in the presence of an internal flow of $v_0 = 300\,m/s$, $v_1 = v_2 = 0$, the riser keeps oscillating. From the frequency domain response of Fig. 4.1(b), the while the response in the presence of an internal flow of $v_0 = 300$ *m/s*, $v_1 = v_2 = 0$, the riser keeps oscillating. From the frequency domain response of Fig. 4.1(b), the natural fundamental frequencies are found to be $\omega = 0.12$ *Hz* for $v_0 = 300$ *m/s*, and $\omega = 0.25$ *Hz* for $v_0 = 0$. This reduction in fundamental frequency with increasing flow velocity is a well known characteristic of such problems, and plays a crucial role in riser stability considerations.

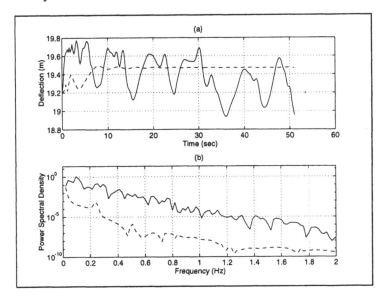

4.1 Free Vibration Response of Top End. (a) Time Domain, (b) Frequency Domain. ($v_0 = 0$ – Dashed line, $v_0 = 300 \ m/s$ – Solid line).

Fig. 4.2 shows the riser at its initial position (solid line) and its position at the end of the simulation, with its curvature a result of its own weight. The dashed line represents the riser with zero internal flow velocity, while the dotted line is for nonzero internal flow. Again, it is seen that in the absence of internal velocity the riser reaches a steady state equilibrium position, while in the presence of internal flow it will keep oscillating due to the flow induced vibrations. The horizontal dashed line represents the mean water level.

Figs. 4.3 and 4.4 depict the response of the riser without and with flow in a three–dimensional rendering of the data. Fig. 4.3 corresponds to the response with no flow of Figs. 4.1(a) and 4.2, where an equilibrium is reached. Fig. 4.4 corresponds to the response with flow of Figs. 4.1(a) and 4.2, where there is a continuous oscillatory motion.

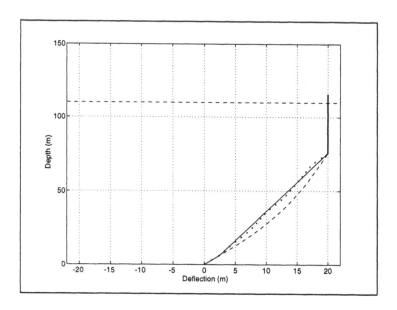

4.2 Riser's Position. Initial state - Solid Line. Final State: Without Flow - Dashed Line, With Flow - Dotted Line.

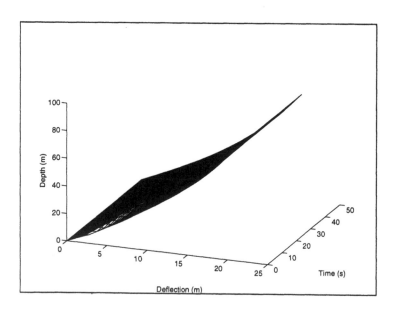

4.3 Response of the Riser with Zero Internal Flow.

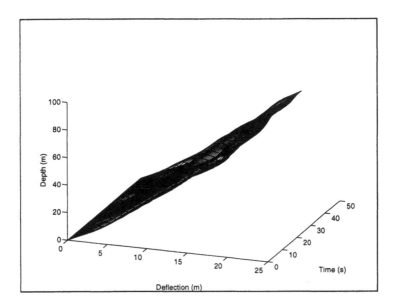

4.4 Response of the Riser with Internal Flow.

Response to Wave and Current

In this section the shifting of the equilibrium position due to current, and the response due to wave excitation, are presented. Fig. 4.5 depicts the equilibrium position of the riser. It is seen that it is the same for both cases of with and without internal flow. This is due to the damping effect of the current as it flows past the riser. This current–induced damping was explained in Chapter 3, Section 3.4.1.

Fig. 4.6 shows the time and frequency domain responses of the riser to wave excitation with wave height of $H = 2\ m$. Again, it is seen that without internal flow, the top end of the riser vibrates with small amplitudes at the frequency of the wave, 0.25 Hz, while with internal flow the amplitude of oscillation is larger. This fact has implications for stability and for design considerations at the interface between the riser and the structure/ship to which it is connected.

The position of the riser is depicted for a particular time in Fig. 4.7. In the absence of internal flow the riser oscillates with small amplitudes about its equilibrium position due to its own weight while with non–zero flow it oscillates about a different equilibrium position and with larger amplitude.

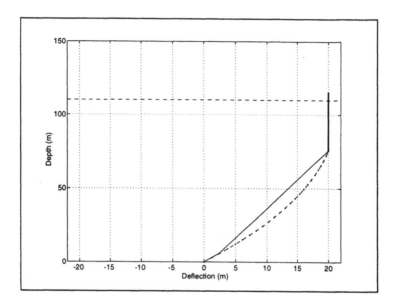

4.5 Equilibrium Position Due to Current. Without Flow - Dashed Line, With Flow - Dotted Line.

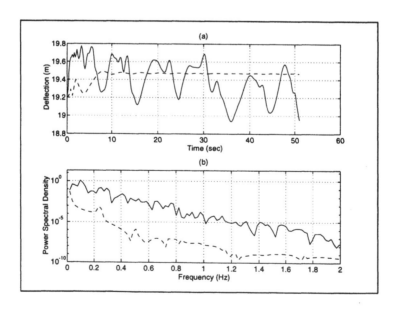

4.6 Response Due to Waves, $H = 2\ m$. (a) Time Domain, (b) Frequency Domain. ($v_0 = 0$ – Dashed line, $v_0 = 300\ m/s$ – Solid line).

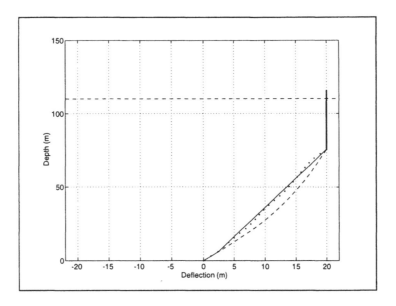

4.7 Response Due to Waves, $H = 2\ m$. With Internal Flow – Dotted Line, Without Internal Flow – Dashed Line.

Response of Riser to Base and Top Excitation

The response to base and top excitation is investigated here. The base is assumed to oscillate at a frequency close to the first natural frequency of the riser with zero internal flow, $\omega_e \approx \omega = 0.25\ Hz$ at $v_0 = 0$, in the x and y directions as follows; $u = 0.2\cos(\omega_e t)\ m$ and $w = 0.1\cos(\omega_e t)\ m$. The response near the top end in the time and frequency domains is shown in Fig. 4.8. The response with internal flow (solid line) is larger than the one without flow (dashed line), although the frequency of excitation is at the first natural frequency of the riser with zero flow. The reason for this counter–intuitive result is that the first natural frequency of the riser with internal flow is about $\frac{1}{2}\omega_e$, and due to the nonlinearities in the system, superharmonics occur. Thus, there is large amplitude response at a superharmonic where more flow energy exists.

The total response of the riser to base excitation is shown in Figs. 4.9 and 4.10. For the later case with internal flow, Fig. 4.10 graphically shows how the flow creates a well-defined regimen for the structural vibration.

As mentioned earlier, the excitation of the top of the riser is embedded into the boundary conditions. Here, it is assumed that the frequency of this excitation is close to the first natural frequency of the riser with zero flow, that is, $y(x = L) = 1\cos(\omega_t t)$, where $\omega_t = \omega = 0.25\ Hz$ at $v_0 = 0$. The response to top excitation in the time and frequency domains is shown in Fig. 4.11. The frequency of excitation coincides with the natural frequency, resulting in high amplitude for both cases of with and without flow. The total response of the riser to top excitation is shown in Figs. 4.12 and 4.13.

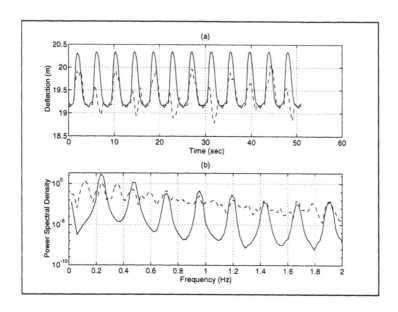

4.8 Response Near to Top End to Base Excitation. (a) Time Domain, (b) Frequency Domain. ($v_0 = 0$ - Dashed line, $v_0 = 300 \ m/s$ - Solid line).

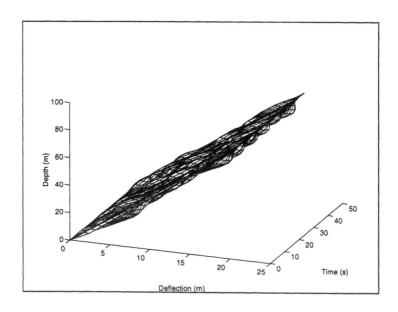

4.9 Response of the Riser to Base Excitation With Zero Internal Flow.

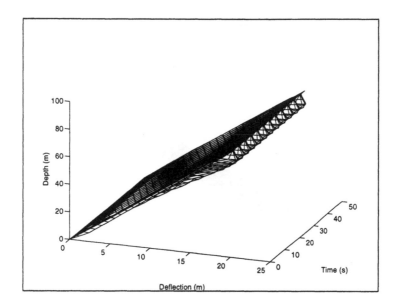

4.10 Response of the Riser to Base Excitation With Internal Flow.

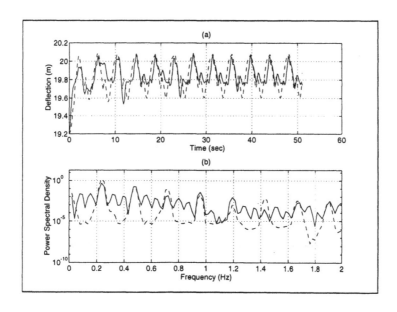

4.11 Response to Top Excitation. (a) Time Domain, (b) Frequency Domain. ($v_0 = 0$ – Dashed line, $v_0 = 300\ m/s$ – Solid line).

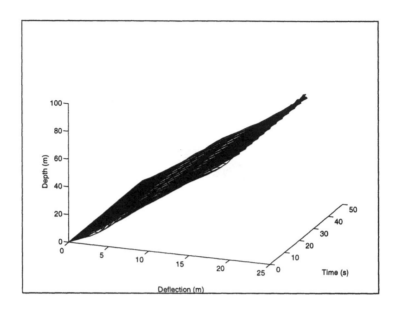

4.12 Response of the Riser to Top Excitation With Zero Internal Flow.

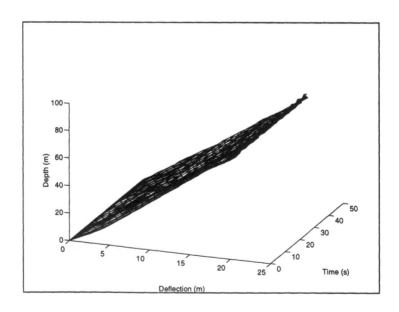

4.13 Response of the Riser to Top Excitation With Internal Flow.

Response to Non–Constant Flow Velocity

The influence of an accelerating flow is now investigated. The internal fluid velocity and acceleration are given in equations 4.4 and 4.5. This is representative of the start-up and shutdown of the pump.

Fig. 4.14 shows the response near the top end of the riser to the oscillating flow, $v_{fl} = 300\cos\omega_{fl}t$ *m/s* and $a_{fl} = -v_2\omega_{fl}\sin\omega_{fl}t$ *m/s*^{2}, where $\omega_{fl} = \omega = 0.25$ *Hz* for $v_0 = 0$. The response is clearly nonlinear, as can be seen from the frequency domain plot, which shows the fundamental frequency and its multiples.

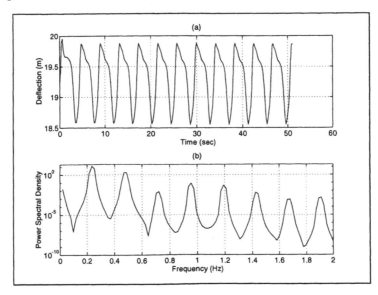

4.14 Response Near the Top End of the Riser to Oscillating Flow. (a) Time Domain, (b) Frequency Domain.

Fig. 4.15 depicts the total response of the riser to oscillating flow.

The equilibrium position of the riser in the presence of positive acceleration (dashed line) and negative acceleration (dotted line) is shown in Fig. 4.16. From the figure it can be seen that a negative acceleration results in a normal catenary but a positive acceleration leads to an inverse catenary shape. This is due to the fact that a negative acceleration is in the direction of the gravitational acceleration, resulting in a 'heavier weight' while positive acceleration is opposite to the direction of gravity, hence is large enough (in this case) to cause a 'buoyancy effect'.

The total response of the riser to a flow having constant positive and negative accelerations is shown in Figs. 4.17 and 4.18, respectively.

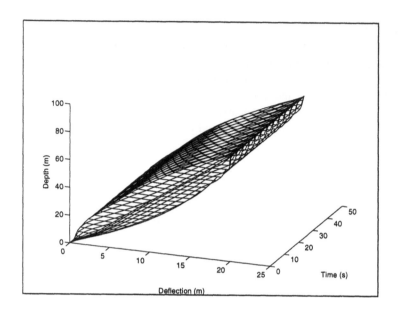

4.15 Response of the Riser to Oscillating Flow.

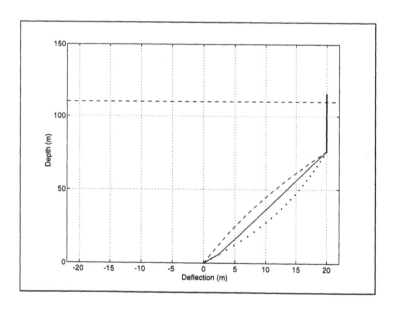

4.16 Equilibrium Position in the Presence of Negative Acceleration (dotted line) and Positive Acceleration (dashed line).

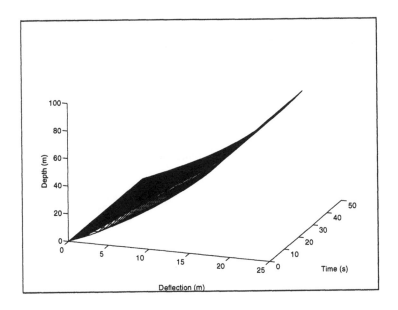

4.17 Response of the Riser to Negative Acceleration.

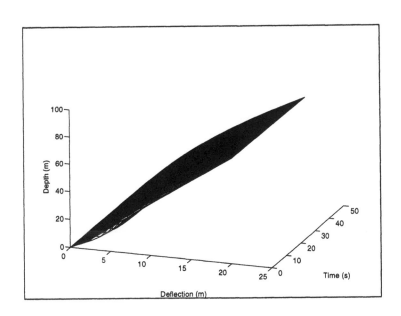

4.18 Response of the Riser to Positive Acceleration.

156

Summary

In this chapter, preliminary studies have been performed for the response of a riser conveying an internal fluid along its length to various ocean and end forces. Of particular importance are the cases where large amplitude vibration occurs at loading frequencies that are sub– and super–harmonics of the structure. Stability is a primary consideration for such systems. These studies begin to map the design domain, and thus provide guidance to the analyst and designer.

The analytical models developed herein can be extended to include greater complexities in loading, structural model, and connection between riser and external structure.

5
Possible Future Studies

Now that this monograph is complete, it is important to provide the reader with some ideas regarding where improvements are possible in the various technical topics developed in the last four chapters.

The possible future work can be divided into two main groups. The first using the models derived in this monograph to conduct more studies, and the second is to refine the models.

Future Studies Using Models Developed in This Work

For all the applications developed in this work, the following additional numerical work is to be expected.

- Earthquake response - The equation of motion includes base excitation that could be due to an earthquake. The response of an articulated tower subjected to earthquake together with waves, wind, and current needs to be investigated.
- Parametric instability analysis - Since the appendaged tower is subjected to vertical base excitation, parametric instability can occur. Mapping the regions of stability and instability due to different amplitudes and frequencies of excitation is important.
- Reliability and fatigue analysis - An expression for the equivalent stress along the tower is derived. This could be used to determine the reliability and fatigue criteria under different environmental conditions.
- Chaos regions - As shown in Chapter 3, due to nonlinearity, chaos regions exist. Mapping these regions with different parameters could be an important and challenging study.

- Application - Using the models derived in this work and applying them to real offshore structures would be important.

Enhancing the Models Developed in this Monograph

Refined models for the structure as well as for the external forces can be formulated;

- Tension Leg Platform with multiple cables, and cables with extension require future study.
- Risers with extension, with non–standard materials, and material nonlinearities pose significant challenges.
- Nonlinear wave theories that include high and steep wave heights can be used in Morison equation.
- Use of a different approximation for the wave forces, rather than Morison's equation can be considered.
- More physically realistic vortex–shedding models will improve the understanding of the coupling between fluid and structure.

In conclusion, the contributions of this monograph are based on the work of many who have expended much time and effort to increase the level of understanding of these problems. These ideas are offered in fully realizing that they offer another view and a few more steps towards a better physical understanding of these structures.

References

[1] A. T. Ippen. *Estuary and Coastline Hydrodynamics*. McGraw-Hill, 1966.

[2] J.R. Morison, M.P. O'Brien, J.W. Johnson, and S.A. Schaaf. The force by surface waves on piles. *Petroleum Transactions, AIME*, 189:149–154, 1950.

[3] L.E. Borgman. Ocean wave simulation for engineering design. *Journal of the Waterways and Harbors Division, ASCE*, 95:557–583, 1969.

[4] J.F. Wilson. *Dynamics of Offshore Structures*. John Wiley and Sons, 1984.

[5] A.W. Dare. *Schaum's Outline of Theory and Problems of Lagrangian Dynamics*. McGraw-Hill, 1967.

[6] P. Bar-Avi and I. Porat. A sound derivation of the nonlinear equations of motion of a traveling string. *The International Journal of Mechanical Engineering Education*, 1994.

[7] W. Weaver, S.P. Timoshenko, and D.H. Young. *Vibration Problems in Engineering*. John Wiley and Sons, 1990.

[8] O.M. Faltinsen. *Sea Loads on Ships and Offshore Structures*. Cambridge University Press, 1994.

[9] M.H. Patel. *Dynamics of Offshore Structures*. Butterworths, 1988.

[10] S.K. Chakrabarti. *Hydrodynamics of Offshore Structures*. Computational Mechanics Publications, 1987.

[11] E. Mitchell. Solving pdes in acsl by dicretization and conversion to odes. *Mitchell and Gauthier Associates*, pages 1–16, 1994.

[12] Y. Jaluria. *Computer Methods for Engineering*. Allyn and Bacon, 1988.

[13] G.E. Burnd and G.C. D'Amorim. Buoyant tower for phase one development of Garoupa field. In *9th Annual Offshore Technology Conference*, pages 177–184, 1977.

[14] D.L. Hays, M. McSwiggan, and R. Vilain. Operation of an articulated oil loading column at the Beryl field in the North Sea. In *11th Annual Offshore Technology Conference*, pages 1805–1815, 1979.

[15] J.R. Smith. The application of a concrete articulating buoyant column to offshore drilling and production systems. In *Proceedings Symposium on New Technology for Exploration and Exploitation of Oil and Gas Resources*, pages 377–389, 1979.

[16] J.R. Smith and R.S. Taylor. The development of articulated buoyant system column system as an aid to economic offshore production. In *European Offshore Petroleum Conference and Exhibition*, pages 545–557, 1980.

[17] H.G. Butt, J. Salewski, and P. Wagner. A large-scale test with the concrete articulated tower CONAT in the vicinity of the research platform 'nordsee'. In *International Conference on Marine Science and Ocean Engineering*, pages 31–47, 1980.

[18] A. Naess. Loads and motions of an articulated loading platform with moored tanker. In *12th Annual Offshore Technology Conference*, pages 409–417, 1980.

[19] P. Bar-Avi and H. Benaroya. Response of an articulated tower to loads due to wave slamming, wind and coriolis acceleration. In *The 36th AIAA/ASME/ASCE/AHS/ASC Structures, Structural Dynamics, And Materials Conference*, pages 257–265, 1995.

[20] Anonymous. BHP brings Challis onstream with world's largest SALRAM. *Ocean Industry*, pages 53–56, 1990.

[21] S.K. Chakrabarti and D.C. Cotter. Analysis of a tower-tanker system. In *Proceedings*

of the 10th Annual Offshore Technology Conference, pages 1301–1310, 1978.

[22] S.K. Chakrabarti and D.C. Cotter. Motion analysis of articulated tower. *Journal of the Waterway, Port, Coastal and Ocean Division, ASCE*, 105:281 – 292, 1979.

[23] C.H. Kim and P.A. Luh. Motions and loads of an articulated loading platform in waves. *IEEE, Ocean Engineering*, pages 956–961, 1981.

[24] P.K. Muhuri and A.S. Gupta. Stochastic stability of tethered buoyant platforms. *Ocean Engineering*, 10(6):471 – 479, 1983.

[25] J.M.T. Thompson, A.R. Bokaian, and R. Ghaffai. Stochastic and chaotic motions of compliant offshore structures and articulated mooring towers. *Journal of Energy Resources Technology*, 106:191 – 198, 1984.

[26] J. Chantrel and P. Marol. Subharmonic response of articulated loading platform. In *Proceedings of the 6th Conference on Offshore Mechanics and Arctic Engineering*, pages 35–45, 1987.

[27] A.K. Jain and T.K. Datta. Stochastic response of articulated towers. In *Deep Offshore Technology 4th International Conference and Exhibit*, pages 191–208, 1987.

[28] T.K. Datta and A.K. Jain. Response of articulated tower platforms to random wind and wave forces. *Computers and Structures*, 34(1):137 – 144, 1990.

[29] A.K. Jain and T.K. Datta. Nonlinear behavior of articulated tower in random sea. *Journal of Engineering for Industry*, 113:238 – 240, 1991.

[30] L.N. Virgin and S.R. Bishop. Catchment regions of multiple dynamic responses in nonlinear problems of offshore mechanics. *Journal of Offshore Mechanics and Arctic Engineering*, pages 127–133, 1990.

[31] H.S. Choi and J.Y.K. Lou. Nonlinear behaviour of an articulated offshore loading platform. *Applied Ocean Research*, 12(2):63 – 74, 1991.

[32] O. Gottlieb, C.S. Yim, and R.T. Hudspeth. Analysis of nonlinear response of an articulated tower. *International Journal of Offshore and Polar Engineering*, 2(1):61 – 66, 1992.

[33] P. Bar-Avi and H. Benaroya. Nonlinear dynamics of an articulated tower submerged in the ocean. *Journal of Sound and Vibration*, 190(1):77–103, 1996.

[34] P. Bar-Avi. *Dynamic response of an offshore articulated tower*. PhD thesis, Rutgers, The State University of New Jersey, 1996.

[35] C.L. Kirk and R.K. Jain. Response of articulated tower to waves and current. *The 9th Annual Offshore Technology Conference*, pages 545 – 552, 1977.

[36] O.A. Olsen, A. Braathen, and A.E. Loken. Slow and high frequency motions and loads of articulated single point mooring system for large tanker. *Norwegian Marine Research*, pages 14–28, 1978.

[37] S.K. Chakrabarti and D.C. Cotter. Transverse motion of articulated tower. *Journal of the Waterway, Port, Coastal and Ocean Division, ASCE*, 107:65 – 77, 1980.

[38] T.E. Schellin and T. Koch. Calculated response of an articulated tower in waves comparison with model tests. In *Proceedings of the Forth Offshore Mechanics and Arctic Engineering Symposium, ASME, Vol. 1*, 1985.

[39] C.Y. Liaw. Bifurcations of subharmonics and chaotic motions of articulated towers. *Engineering Structures*, pages 117–124, 1988.

[40] C.W. Liaw, N.J. Shankar, and K.S. Chua. Large motion analysis of compliant structures using Euler parameters. *Ocean Engineering*, 16(6):545–557, 1989.

[41] J.W. Leonard and R.A. Young. Coupled response of compliant offshore platforms. *Engineering Structures*, 7:74 – 84, 1985.

[42] C.Y. Liaw, N.J. Shankar, and K.S. Chua. Subharmonic motions and wave force-structure interaction. *Marine Structures*, 5:281–295, 1992.

[43] P. Bar-Avi and H. Benaroya. Response of a two DOF articulated tower to different environmental conditions. *International Journal of Nonlinear Mechanics*, 1995.

[44] P. Bar-Avi and H. Benaroya. Dynamic response of an articulated tower to random waves and current loads. In *Third International Conference on Stochastic Structural Dynamics*, Puerto Rico, (1995), to be published.

[45] P. Bar-Avi and H. Benaroya. Stochastic response of an articulated tower. *International Journal of Nonlinear Mechanics*, 1995.

[46] R.K. Jain and C.L. Kirk. Dynamic response of a double articulated offshore loading structure to noncollinear waves and current. *Journal of Energy Resources Technology*, 103:41 – 47, 1981.

[47] L.L. Seller and J.M. Niedzwecki. Response characteristics of multi-articulated off-shore towers. *Ocean Engineering*, 19(1):1 – 20, 1992.

[48] K.F. Haverty, J.F. McNamara, and B. Moran. Finite dynamic motions of articulated offshore loading towers. In *Proceedings of the Internation Conference of Marine Research Ship Technology and Ocean Engineering*, 1982.

[49] J.F. McNamara and M. Lane. Practical modeling for articulated risers and loading columns. *Journal of Energy Resources Technology*, pages 444–450, 1984.

[50] G. Sebastiani, R. Brandi, F. D. Lena, and A. Nista. Highly compliant column for tanker mooring and oil production in 1000 m water depth. In *The 16th Annual Offshore Technology Conference*, pages 379–388, 1984.

[51] S.Y. Hanna, D.I. Karsen, and J.Y. Yeung. Dynamic response of a compliant tower with multiple articulations. In *Seventh International Conference on Offshore Mechanics and Arctic Engineering*, pages 257–269, 1988.

[52] I.H. Hevacioglu and A. Incecik. Dynamic analysis of coupled articulated tower and floating production system. In *The 7th International Conference on Offshore Mechanics and Arctic Engineering*, pages 279–287, 1988.

[53] K. Yoshida, H. Suzuki, and N. Ona. Control of dynamic response of tower-like offshore structures in waves. In *7th International Conference on Offshore Mechanics and Arctic Engineering, ASME*, volume 1, pages 249–256, 1988.

[54] K. Yoshida, H. Suzuki, and S. Mishima. Response control of articulated compliant tower. In *Eighth International Conference on Offshore Mechanics and Arctic Engineering*, pages 199–207, 1989.

[55] C. Ganapathy, B.P. Samantaray, and K. Nagamani. Truss type articulated towers for deep water applications. In *The 9th International Conference on Offshore Mechanics and Arctic Engineering*, pages 477–484, 1990.

[56] K.M. Mathisen and P.G. Bergan. Large displacement analysis of submerged multibody systems. *Engineering Computations*, 9:609–634, 1992.

[57] M. Issacson. Wave and current forces on fixed offshore structure. *Canadian Journal Civil Engineering*, 15:937 – 947, 1988.

[58] K.Y.R. Billah. *A Study of Vortex-Induced Vibration*. PhD thesis, Princeton University, 1989.

162

[59] Y. Dong and J.Y.K. Lou. Vortex-induced nonlinear oscillation of tension leg platform tethers. *Ocean Engineering*, 18(5):451–464, 1991.

[60] S.K. Chakrabarti. *Nonlinear Methods in Offshore Engineering*. Elsevier Science Publishing Company Inc., 1990.

[61] J.P. Hooft. *Advanced Dynamics of Marine Structures*. John Wiley and Sons, 1982.

[62] N. Hogben, B.L. Miller, J.W. Searle, and G. Ward. Estimation of fluid loading on offshore structures. *Proceeding of Civil Engineering*, pages 515–562, 1977.

[63] R. Adrezin, P. Bar-Avi, and H. Benaroya. Dynamic response of compliant offshore structure. *ASCE, J. Aerospace Engineering*, 1996.

[64] B.J. Natvig and P. Teigen. Review of hydrodynamic challenges in TLP design. *International Journal of Offshore and Polar Engineers*, 3(4):241–249, December 1993.

[65] S.K. Chakrabarti. *Hydrodynamics of Offshore Structures*. Computational Mechanics Publications, 1987.

[66] A. Salpukas. Oil companies drawn to the deep. *The New York Times*, pages D1,D5, December 7 1994.

[67] R. Robison. Bullwinkle's big brother. *Civil Engineering*, pages 44–47, 1995.

[68] M. Sato, T. Natsume, N. Kodan, K. Ishikawa, S. Ito, and K. Tono. NKK TLP (tension leg platform). *NKK Technical Review*, (55):103–114, 1989.

[69] J.W. Pallini, Jr. and A. Yu. TLP tendon connections top and bottom terminations and tendon pipe couplings. *Offshore and Arctic Operations*, 51:111–118, 1993.

[70] F.K. Lim and S. Hatton. Design considerations for TLP risers in harsh environments. *Proceedings of the First International Offshore and Polar Engineering Conference*, 2:182–189, August 1991.

[71] P.A. Abbott, R.B. D'Souza, I.C. Solberg, and K. Eriksen. Evaluating deepwater development concepts. *SPE International Petroleum Conference and Exhibition of Mexico*, pages 111–127, Oct 1994.

[72] B.W. Oppenheim and J.E. Fletter. Design notes on spread mooring systems. *Proceedings of the First International Offshore and Polar Engineering Conference*, 2:259–265, August 1991.

[73] P.G.S. Dove and C.J. Lohr. Lateral mooring systems for tension leg platforms. *Offshore and Arctic Operations*, 51:129–138, 1993.

[74] K. Yoshida, H. Suzuki, D. Nam, M. Hineno, and S. Ishida. Active control of coupled dynamic response of TLP hull and tendon. *Proceedings of the Fourth International Offshore and Polar Engineering Conference*, 1:98–104, 1994.

[75] R.G. Bea, C.A. Cornell, J.E. Vinnem, J.F. Geyer, G.J. Shoup, and B. Stahl. Comparative risk assessment of alternative TLP systems: Structure and foundation aspects. *Journal of Offshore Mechanics and Arctic Engineering*, 116:86–96, May 1994.

[76] P.G. Wybro. Advances in methods for deepwater tlp installations. *Offshore and Arctic Operations*, 68:201–212, 1995.

[77] R.C.T. Rainey. The dynamics of tethered platforms. *Meeting of the Royal Institution of Naval Architects*, pages 59–80, April 1977.

[78] D.T. Armenis, T.A. Angelopoulos, and D.G. Papanikas. Time domain simulation of the dynamic behaviour of a tension leg platform. *Proceedings of the First International Offshore and Polar Engineering Conference*, 1:100–107, August 1991.

[79] B.B. Mekha, C.P. Johnson, and J.M. Roesset. Effects of different wave free surface

approximations on the response of a TLP in deep water. *Proceedings of the Fourth International Offshore and Polar Engineering Conference*, 1:105–115, 1994.

[80] X. Song and A. Kareem. Combined system analysis of tension leg platforms: A parallel computational scheme. *OMAE*, 1:123–134, 1994.

[81] C.P. Johnson. A computer aided design approach for deep water tension leg platforms. *Offshore and Arctic Operations*, 58:77–87, 1994.

[82] C.H. Kim, M.H. Kim, Y.H. Liu, and C.T. Zhao. Time domain simulation of nonlinear response of a coupled TLP system in random seas. *Proceedings of the Fourth International Offshore and Polar Engineering Conference*, 1:68–77, 1994.

[83] Y.H. Liu, M.H. Kim, and C.H. Kim. The computation of second-order mean and double-frequency wave loads on a compliant tlp by HOBEM. *International Journal of Offshore and Polar Engineering*, 5(2):111–119, Jun 1995.

[84] A. Kareem. Nonlinear dynamic analysis of compliant offshore platforms subjected to fluctuating wind. *Journal of Wind Engineering and Industrial Aerodynamics*, 14:345–356, 1983.

[85] A. Kareem. Wind-induced response analysis of tension leg platforms. *Journal of Structural Engineering*, 111(1):37–55, January 1985.

[86] A. Kareem, P.C. Lu, T.D. Finnigan, and S.L.V. Liu. A wind tunnel investigation of aerodynamic loads on a typical tension leg platform. *Offshore Technology Conference*, pages 187–197, May 1986.

[87] Y. Li and A. Kareem. Computation of wind-induced drift forces introduced by displaced position of compliant offshore platforms. *Journal of Offshore Mechanics and Arctic Engineering*, 114:175–184, August 1992.

[88] A. Kareem and Y. Li. Response of tension leg platforms to wind, wave, and currents: A frequency domain approach. *Offshore Technology Conference*, pages 437–442, May 1990.

[89] A. Kareem and Y. Li. Wind-excited surge response of tension-leg platform: Frequency-domain approach. *Journal of Engineering Mechanics*, 119(1):161–173, January 1993.

[90] A. Kareem and J. Zhao. Response statistics of tension leg platforms under wind and wave loads: A statistical quadratization approach. *Structural Safety and Reliability*, pages 497–503, 1994.

[91] A. Kareem and J. Zhao. Analysis of non-gaussian surge response of tenion leg platforms under wind loads. *Journal of Offshore Mechanics and Arctic Engineering*, 116:137–144, August 1994.

[92] Y. Li and A. Kareem. Parametric modelling of stochastic wave effects on offshore platforms. *Applied Ocean Research*, 15:63–83, 1993.

[93] A. Kareem and Y. Li. Stochastic response of a tension leg platform to viscous and potential drift forces. *Probabilistic Engineering Mechanics*, 9:1–14, 1994.

[94] Y. Li and A. Kareem. Stochastic decomposition and application to probabilistic dynamics. *Journal of Engineering Mechanics*, 121(1):162–174, January 1995.

[95] P.J. Vickery. Wind-induced response of tension leg platform: Theory and experiment. *Journal of Structural Engineering*, 121(4):651–663, April 1995.

[96] E.R. Jefferys and M.H. Patel. On the dynamics of taut mooring systems. *Engineering Structures*, 4:37–43, Jan 1982.

[97] M.H. Patel and E.J. Lynch. Coupled dynamics of tensioned buoyant platforms and

mooring tethers. *Engineering Structures*, 5:299–308, October 1983.

[98] M.H. Patel and F.B. Seyed. Review of flexible riser modelling and analysis techniques. *Enginnering Structures*, pages 293–304, 1994.

[99] C.P. Lee and J.F. Lee. Wave-induced surge motion of a tension leg structure. *Ocean Engineering*, 20(2):171–186, 1993.

[100] T.P. Mullarkey and J.F. McNamara. Evaluation of the three-dimensional linear hydro-dynamics of the issc TLP using a semianalytical method. *Proceedings of the Fourth International Offshore and Polar Engineering Conference*, 1:61–67, 1994.

[101] W.S. Park, C.B. Yun, and B.K. Yu. Reliability analysis of tension leg platforms by domain crossing approach. *Proceedings of the First International Offshore and Polar Engineering Conference*, 1:108–116, August 1991.

[102] G.S. Liou, J. Penzien, and R.W. Yeung. Response of tension-leg platforms to vertical seismic excitations. *Earthquake Engineering and Structural Dynamics*, 16:157–182, 1988.

[103] T. Kawanishi, T. Kato, H. Takamura, and H. Kobayashi. Earthquake response of the tension leg platform under offset condition. *Proceedings of the First International Offshore and Polar Engineering Conference*, 1:80–86, 1991.

[104] T. Kawanishi, S. Ohashi, H. Han-Ya, and H. Kobayashi. Tsunami response of the tension leg platform under unbalanced initial tension. *Oceans Conference Record*, 2:42–47, 1994.

[105] K. Venkataramana. Earthquake response of tension-leg-platforms in steady currents. *Earthquake Engineering and Structural Dynamics*, 23:63–74, 1994.

[106] T. Munkejord. Heidrun concrete tlp: Update. *Offshore and Arctic Operations*, 68:189–196, 1995.

[107] S.B. Kim, E.J. Powers, R.W. Miksad, and F.J. Fischer. Quantification of nonlinear energy transfer to sum and difference frequency responses of TLP's. *Proceedings of the First International Offshore and Polar Engineering Conference*, 1:87–92, August 1991.

[108] B.J. Natvig and H. Vogel. Sum-frequency excitations in TLP design. *Proceedings of the First International Offshore and Polar Engineering Conference*, 1:93–99, August 1991.

[109] X.B. Chen, B. Molin, and F. Petitjean. Numerical evaluations of the springing loads on tension leg platforms. *Marine Structures*, 8:501–524, 1995.

[110] T. Marthinsen and J. Muren. Snorre TLP - analysis of measured response. *Proceedings of the Fourth International Offshore and Polar Engineering Conference*, 1:32–39, 1994.

[111] B.J. Natvig. A proposed ringing analysis model for higher order tether response. *Proceedings of the Fourth International Offshore and Polar Engineering Conference*, 1:40–51, 1994.

[112] K.R.Gurley and A. Kareem. Numerical experiments in ringing and springing of offshore platforms. *Proceedings of the Tenth ASCE Eng. Mech. Conference*, pages 1–4, May 1995.

[113] J.S. Chung. Added mass and damping on an oscillating surface-piercing circular column with circular footing. *Proceedings of the Fourth International Offshore and Polar Engineering Conference*, 3:182–189, April 1994.

[114] D.C. Weggel and J.M. Roesset. Vertical hydrodynamic forces on truncated cylinders. *Proceedings of the Fourth International Offshore and Polar Engineering Conference*, 3:210–217, April 1994.

[115] R. Gopalkrishnan, M.S. Triantafyllou, and M.A. Grosenbaug. Influence of amplitude modulation of the fluid forces acting on a vibrating cylinder in cross-flow. *Proceedings of the First International Offshore and Polar Engineering Conference*, 2:132–139, August 1991.

[116] M. Hogedal, J. Skourup, and H.F. Burcharth. Wave forces on a vertical smooth cylinder in directional waves. *Proceedings of the Fourth International Offshore and Polar Engineering Conference*, 3:218–224, April 1994.

[117] K. Hoshino and H. Sato. Experimental study on hydrodynamic forces acting on an oscillating column with circular footing. *Proceedings of the Fourth International Offshore and Polar Engineering Conference*, 3:173–181, April 1994.

[118] Y.G. Lee, S.W. Hong, and K.J. Kang. A numerical simulation of vortex motion behind a circular cylinder above a horizontal plane boundary. *Proceedings of the Fourth International Offshore and Polar Engineering Conference*, 3:427–433, April 1994.

[119] G. Moe, K. Holden, and P.O. Yttervoll. Motion of spring supported cylinders in subcritical and critical water flows. *Proceedings of the Fourth International Offshore and Polar Engineering Conference*, 3:468–475, April 1994.

[120] I.A. Sibetheros, R.W. Miksad, A.V. Ventre, and K.F. Lambrakos. Flow mapping on the reversing vortex wake of a cylinder in planar harmonic flow. *Proceedings of the Fourth International Offshore and Polar Engineering Conference*, 3:406–412, April 1994.

[121] B.M. Sumer and A. Kozakiewicz. Visualization of flow around cylinders in irregular waves. *Proceedings of the Fourth International Offshore and Polar Engineering Conference*, 3:413–420, April 1994.

[122] S. Sunahara and T. Kinoshita. Flow around circular cylinder oscillating at low Keulegan-Carpenter number. *Proceedings of the Fourth International Offshore and Polar Engineering Conference*, 3:476–483, April 1994.

[123] J.S. Chung, A.K. Whitney, D. Lezius, and R.J. Conti. Flow-induced torsional moment and vortex suppression for a circular cylinder with cables. *Proceedings of the Fourth International Offshore and Polar Engineering Conference*, 3:447–459, April 1994.

[124] J.M. Niedzwecki and J.R. Huston. Wave interaction with tension leg platforms. *Ocean Engineering*, 19(1):21–37, 1992.

[125] J.M. Niedzwecki and O. Rijken. Characterizing some aspects of stochastic wave-structure interactions. *Stochastic Dynamics and Reliability of Nonlinear Ocean Systems*, 77:109–118, 1994.

[126] A.S. Duggal and J.M. Niedzwecki. Dynamic response of a single flexible cylinders in waves. *Journal of Offshore Mechanics and Arctic Engineering*, 117:99–104, 1995.

[127] O.M. Faltinsen, J.N. Newman, and T. Vinje. Nonlinear wave loads on a slender vertical cylinder. *J. Fluid Mech.*, 289:179–198, 1995.

[128] M-H. Kim. Interaction of waves with n vertical circular cylinders. *Journal of Waterway, Port, Coastal, and Ocean Engineering*, 119(6):671–689, 1993.

[129] S. Zhu, Y. Cai, and S.S. Chen. Experimental fluid-force coefficients for wake-induced cylinder vibration. *Journal of Engineering Mechanics*, pages 1003–1015, Sept 1995.

166

[130] N. Mizutani, C. Kim, K. Iwata, R. Matsudaira, Y. Miyaike, and H. Yu. Wave forces acting on multiple cylindrical structures with large diameter. *Proceedings of the Fourth International Offshore and Polar Engineering Conference*, 3:244–251, April 1994.

[131] G. Moe, K.C. Stromsem, and I. Fylling. Behaviour of risers with internal flow under various boundary conditions. *Proceedings of the Fourth International Offshore and Polar Engineering Conference*, 2:258–262, April 1994.

[132] M.H. Patel and F.B. Seyed. Review of flexible riser modelling and analysis techniques. *Engineering Structures*, 17(4):293–304, 1995.

[133] A. Bokaian. Lock-in prediction of marine risers and tethers. *Journal of Sound and Vibration*, 175(5):607–623, 1994.

[134] Y.C. Kim and P.M. Lee. Nonlinear motion characteristics of a long vertical cylinder. *Proceedings of the First Pacific/Asia Offshore Mechanics Symposium*, pages 247–256, June 1990.

[135] S. Saevik and S. Berge. Correlation between theoretical predictions and testing of two 4-inch flexible pipes. *Offshore and Arctic Operations*, 51:63–78, 1993.

[136] S. Chucheepsakul, S. Buncharoen, and C.M. Wang. Large deflection of beams under moment gradient. *Journal of Engineering Mechanics*, 120(9):1848–1860, September 1994.

[137] H.H. Yoo, R.R. Ryan, and R.A. Scott. Dynamics of flexible beams undergoing overall motions. *Journal of Sound and Vibration*, 181(2):261–278, 1995.

[138] M. Iura and S.N. Atluri. Dynamic analysis of planar flexible beams with finite rotations by using inertial and rotating frames. *Computers and Structures*, 55(3):453–462, 1995.

[139] S. Chucheepsakul, S. Buncharoen, and T. Huang. Elastica of simple variable-arc-length beam subjected to end moment. *Journal of Engineering Mechanics*, 121(7):767–772, July 1995.

[140] F. Benedettini and G. Rega. Non-linear dynamics of an elastic cable under planar excitation. *International Journal of Non-Linear Mechanics*, 22(6):497–509, 1987.

[141] K. Takahashi and Y. Konishi. Non-linear vibrations of cables in three dimensions, part i: Non-linear free vibrations. *Journal of Sound and Vibration*, 118(1):69–84, 1987.

[142] K. Takahashi and Y. Konishi. Non-linear vibrations of cables in three dimensions, part ii: Out-of-plane vibrations under in-plane sinusoidally time-varying load. *Journal of Sound and Vibration*, 118(1):85–97, 1987.

[143] F. Benedettini and G. Rega. Planar non-linear oscillations of elastic cables under superharmonic resonance conditions. *Journal of Sound and Vibration*, 132(3):353–366, 1989.

[144] S.P. Cheng and N.C. Perkins. Free vibration of a sagged cable supporting a discrete mass. *J. Acoust. Soc. Am.*, 91(5):2654–2662, May 1992.

[145] N.C. Perkins. Planar and non-planar response of a suspended cable driven by small support oscillations. *Proceedings of the First International Offshore and Polar Engineering Conference*, 2:210–215, August 1991.

[146] C.T. Howell. Numerical analysis of 2-d nonlinear cable equations with applications to low-tension problems. *Proceedings of the First International Offshore and Polar Engineering Conference*, 2:203–209, August 1991.

[147] M. Irvine. *Cable Structures*. Dover Publications, 1992.

[148] S.P. Cheng and N.C. Perkins. Closed-form vibration analysis of sagged cable/mass

suspensions. *Journal of Applied Mechanics*, 59:923–928, December 1992.

[149] S.P. Cheng and N.C. Perkins. Theoretical and experimental analysis of the forced response of sagged cable/mass suspensions. *Journal of Applied Mechanics*, 61:944–948, December 1994.

[150] F. Benedettini, G. Rega, and R. Alaggio. Non-linear oscillations of a four-degree-of-freedom model of a suspended cable under multiple internal resonance conditions. *Journal of Sound and Vibration*, 182(5):775–798, 1995.

[151] H. Shin. Analysis of extreme tensions in a snapping cable. *Proceedings of the First International Offshore and Polar Engineering Conference*, 2:216–221, August 1991.

[152] M.S. Triantafyllou and C.T. Howell. Dynamic response of cables under negative tension: An ill-posed problem. *Journal of Sound and Vibration*, 173(4):433–446, 1994.

[153] F.B. Seyed and M.H. Patel. Parametric studies of flexible risers. In *The First International Offshore and Polar Engineering Conference*, 1991.

[154] B.B. Mekha, C.P. Johnson, and J.M. Roesset. Implications of tendon modeling on nonlinear response of TLP. *Journal of Structural Engineering*, 122(2):142–149, Feb 1996.

[155] M.S. Triantafyllou. Dynamics of cables and chains. *Shock and Vibration Digest*, pages 3–5, 1994.

[156] G. Moe. Behaviour of risers with internal flow under various boundary conditions. In *Fourth International Offshore and Polar Engineering Conference*, 1994.

[157] K. Aso, K. Kan, H. Doki, and K. Iwato. The effects of vibration absorbers on the longitudinal vibration of a pipe string in the deep sea. In *The First International Offshore and Polar Engineering Conference*, 1991.

[158] A. McCone Jr. Technical challanges in the design of flexible pipes. *Offshore and Arctic Operation - ASME*, pages 1–3, 1993.

Index